Properties and Applications of Alginates

Edited by Irem Deniz, Esra Imamoglu and Tugba Keskin-Gundogdu

Published in London, United Kingdom

IntechOpen

Supporting open minds since 2005

Properties and Applications of Alginates
http://dx.doi.org/10.5772/intechopen.94635
Edited by Irem Deniz, Esra Imamoglu and Tugba Keskin-Gundogdu

Contributors
Lei Nie, Yaling Deng, Ningning Yang, Oseweuba Valentine Okoro, Amin Shavandi, Garba M. Khalid, Bairagi C Mallick, Sulata Sahu, Ivana Lourenço de Mello Ferreira, Rodrigo Ferreira Bittencourt, Clenilson Sousa Júnior, S. Giridhar Reddy, Shivangi Omer, Ghaidaa Alkhayer, Esra Imamoglu, Cagla Yarkent, Bahar Aslanbay Guler, Ceren Gurlek, Yaprak Sahin, Ayse Kose, Suphi S. Oncel, Francesca Selmin

First published in London, United Kingdom, 2022 by IntechOpen
IntechOpen is the global imprint of INTECHOPEN LIMITED, registered in England and Wales, registration number: 11086078, 5 Princes Gate Court, London, SW7 2QJ, United Kingdom
Printed in Croatia

British Library Cataloguing-in-Publication Data
A catalogue record for this book is available from the British Library

Additional hard and PDF copies can be obtained from orders@intechopen.com

Properties and Applications of Alginates
Edited by Irem Deniz, Esra Imamoglu and Tugba Keskin-Gundogdu
p. cm.
Print ISBN 978-1-83969-499-8
Online ISBN 978-1-83969-500-1
eBook (PDF) ISBN 978-1-83969-501-8

For EU product safety concerns:
IN TECH d.o.o., Prolaz Marije Krucifikse Kozulić 3, 51000 Rijeka, Croatia, info@intechopen.com or visit our website at intechopen.com.

Meet the editors

Dr. Irem Deniz currently works at the Department of Bioengineering, Manisa Celal Bayar University, Turkey. Her research interests include bioprocesses, enzyme technology, scale-up, bioreactors, gasification, biochemical engineering, bioengineering, bioconversion of agricultural raw and waste commodities into value-added products, bioenergy/biofuels, microalgae, novel bioreactors, and downstream processing. She is an editorial board meeting for *Biochemical Engineering Journal* and a member of the International Bioprocess Association (IBA). To date, she has published more than thirty SCI articles and book chapters.

Dr. Esra Imamoglu is an associate professor in the Bioengineering Department, Ege University, Izmir, Turkey. She obtained a Ph.D. from Ege University in 2011. Her research interests include bioprocess engineering, biofuels, production techniques of microalgae, optimization and modelling, photobioreactor design, scale-up, and CFD simulation. She has been involved in EU activities (EU FP6-FP7 and COST projects) as a partner. To date, she has published more than forty SCI articles and international articles and worked as a reviewer for more than fifteen journals. Additionally, Dr. Imamoglu has organized many national and international scientific meetings.

Dr. Keskin-Gundogdu is an associate professor in the Department of Environmental Protection, Izmir Democracy University, Turkey. She received her BSc in Chemical Engineering from Hacettepe University, Turkey, and an MSc and Ph.D. in Bioengineering from Ege University, Turkey. Her area of expertise is environmental biotechnology with a focus on bioconversion of organic wastes into biofuels (biogas, biohydrogen, and bioethanol), bioreactor configurations, advanced oxidation processes, and biodesign. Throughout her career, she has worked as a researcher in many projects and collaborated with different groups from various countries. She has published several scientific papers, books, and conference proceedings.

Contents

Preface

Alginates were first used in 1883 by researcher Edward Stanford, and commercial development commenced in 1927. Alginate is a seaweed product that is a polysaccharide or carbohydrate polymer with unique properties such as biodegradability, biocompatibility, renewability, and lack of toxicity. These properties give alginates potential numerous applications in biomedical science and engineering, agriculture, drug delivery, enzyme immobilization, heavy metal removal from industrial effluents, as functional food ingredients, mineralization of organic pollution, emulsifiers, consistency enhancers, and thickening agents in cosmetic formulas. However, there are certain limitations to the industrial application of alginate. The most important of these limitations is its exponentially increasing cost with growing scale. The global production of alginates has now risen to about 40,000 tonnes per year.

Many books and papers have addressed important aspects of alginates over the years, helping to disseminate knowledge of these polymers. As alginates continue to grow and many new colleagues continue to be interested in alginate applications, a comprehensive book considering the cutting-edge technologies is a worthwhile contribution. This book presents the current knowledge and recent advances in the field of alginates to stimulate further research. It also includes both fundamental principles and practical applications for professionals and beginners.

This book is organized into eight chapters covering information on alginates and their applications. The chapters are as follows:

Chapter 1 "Nanomagnetic Polymeric Absorbent Based on Alginate and Gamma-Maghemite Synthesized In Situ for Wastewater Treatment from Metallurgical Industry"

Chapter 2 "Alginates - A Seaweed Product: Its Properties and Applications"

Chapter 3 "Heavy Metal Removal by Alginate Based Agriculture and Industrial Waste Nanocomposites"

Chapter 4 "Alginate Metal Complexes and Their Application"

Chapter 5 "Algal Alginate in Biotechnology: Biosynthesis and Applications"

Chapter 6 "Alginate-Based Composite and Its Biomedical Applications"

Chapter 7 "Applications of Alginates in the Design and Preparation of Orodispersible Dosage Forms"

Chapter 8 "Curcumin-Alginate Mixed Nanocomposite: An Evolving Therapy for Wound Healing"

We thank all the authors for their valuable contributions. We are also grateful to our families who were extremely supportive during the editing of this book. Finally, we would like to express our appreciation to Author Service Manager Marica Novakovic at IntechOpen for her guidance during the publication of this book.

Irem Deniz
Manisa Celal Bayar University,
Manisa, Turkey

Esra Imamoglu
Ege University,
Izmir, Turkey

Tugba Keskin-Gundogdu
Demokrasi University,
Izmir, Turkey

Section 1

Applications

Chapter 1

Nanomagnetic Polymeric Absorbent Based on Alginate and Gamma-Maghemite Synthesized In Situ for Wastewater Treatment from Metallurgical Industry

Ivana Lourenço de Mello Ferreira,
Rodrigo Ferreira Bittencourt and Clenilson Sousa Júnior

Abstract

A nanomagnetic absorbent based on calcium alginate was produced successfully with the maghemite nanoparticles synthesized in situ, i.e., together with the polysaccharide crosslinking reaction. Physicochemical properties of the absorbent were analyzed and its ability to remove Ni(II) and Mn(II) ions from a real metallurgical industry wastewater was evaluated. Kinetic studies of the adsorption of these heavy metals were realized. To ascertain the most suitable quantity of absorbent to remove Ni(II) and Mn(II) from the wastewater, the absorbent mass was varied and adsorption kinetics was also evaluated. The competitiveness between the metals was evaluated to understand the adsorption mechanism. The samples were characterized by transmission electron microscopy, vibrating sample magnetometry, X-ray diffractometry and Mössbauer spectroscopy. The absorbent prepared, in this work, can be classified as a hydrogel. It presented predominant spherical morphology and micrometric dimension, containing atoms of iron and calcium dispersed uniformly in their internal and external surfaces. The synthesized maghemite nanoparticles presented superparamagnetic behavior. Results showed that the adsorption equilibrium time for both ions was about 60 min. The removal percentages from wastewater were 60.5% for nickel and 56.6% for manganese, using 300 mg of hydrogel. Results revealed that the adsorption mechanism is by ionic change between calcium and heavy metals.

Keywords: Nanomagnetic absorbent, hydrogel, calcium alginate, gamma-maghemite, adsorption mechanism, metallurgical wastewater

1. Introduction

Environmental pollution caused by heavy metals is a serious problem and has been the focus of worldwide concern. The chemical effect of these metals has been of great environmental interest due the fact these metals are not biodegradable, which means they have cumulative effects in organisms, causing serious damage to health [1, 2]. Among several metals, manganese and nickel are considered highly toxic. Both are contained in wastewater from galvanoplasty as a consequence of the electrodeposition processes and other metallic surface treatments to prevent corrosion [3]. The main harmful effects caused by these metals are cancer, pulmonary lesions and central nervous system damages [4].

There are several methods for the removal of ions from aqueous solutions, such as reverse osmosis, ion exchange, precipitation, electrodialysis and adsorption [5]. Among these methods, adsorption is the most versatile and widely used to remove various pollutants [6]. Methods to remove heavy metals using low-cost adsorbents have been successfully developed, as shown in the literature [7–9].

Natural polymers like alginate (polyanion) are receiving growing attention due to their strong affinity for heavy metal ions. The absorption capacity, specifically of hydrogel in particle form, is comparable to or even better than commercial ion exchange resins. Furthermore, these materials are abundant, biocompatible and environmentally friendly, which make them potential adsorbents for removal of pollutants from wastewater [10].

Accordingly, economically feasible materials that provide efficient removal of these metals from wastewater are being widely studied. The adsorption described in this article is used as a cost-effective way to remove heavy metals, making it competitive with the conventional technology. The proposed adsorbent, consisting of hydrogel based on alginate, a natural anionic polymer, is widely studied to remove heavy metals, since it is nontoxic, inexpensive and highly efficient regarding adsorption [2].

Alginate is a linear copolymer composed of β-D-mannuronic acid (M) and α-L-guluronic acid (G) subunits. It can be used in several fields, like the food, pharmaceutical biomedical industries, for purposes such as drug delivery, sensorial enzyme encapsulation and development of contrast agents for diagnostic imaging. Moreover, alginates have been proven to be outstanding for water purification. Research has also demonstrated that gel beads of calcium alginate can remove heavy metals from wastewater [11].

Nevertheless, the separation of the loaded biomaterials from the medium is often a problem. So, the use of magnetic adsorbents (called here magsorbents) to solve this technical problem has received considerable attention in recent years [12].

Magnetic nanoparticles are embedded in polysaccharides to raise their capacity as biosorbents. Also, they are very useful in the isolation or recovery process of gel beads. Papers have been published describing the removal of heavy metal ions using maghemite and alginate, as magnetic and encapsulating material, respectively [13].

Idris et al. [2] prepared magnetic alginate beads by insertion of nanoparticles of maghemite in chains of sodium alginate, aiming to remove ions of Pb (II) from aqueous solutions. Physicochemical parameters, such as pH, initial metal concentration and contact time were studied. The results revealed that 95.2% of Pb (II) was removed in 2 hours at pH 7, and the maximum adsorption capacity was 50 mg g^{-1}. Furthermore, the presence of magnetic particles in hydrogel beads enabled the easy isolation of the beads after the sorption process. The results showed the potential of using magnetic beads to treat wastewater containing heavy metals.

However, to the best of our knowledge, not studies have been published about the synthesis of hydrogels based on calcium alginate along with γ-maghemite nanoparticles prepared in situ (preparation method in a single step) to remove heavy metals from industrial wastewater. The existing works first synthesize the gamma-maghemite by the co-precipitation process and then add it to the hydrogel (two-step preparation method). Moreover, in our study, the adsorption kinetic in function of hydrogel mass and the competitiveness between the metals (adsorption mechanism) were evaluated.

2. Experimental

All chemicals were bought in analytical purity and were used without further purification. The main chemicals, calcium chloride and ferrous sulfate, were purchased from Vetec. The manganese and nickel solutions (1000 mg L^{-1}) were prepared and used as stock solutions.

Sodium alginate was supplied by Sigma-Aldrich Chemicals Co. The industrial wastewater used in this work was provided by a large metallurgical company, more particularly from the E-coat painting line pretreatment, which is a water-based immersion coating process for application on metal parts.

2.1 Synthesis of nanomagnetic absorbent

The nanomagnetic absorbent were prepared according to the method described in Llanes et al. [14] with some adaptations. It was synthesized an iron oxide-alginate nanocomposites in situ, as follows. A aqueous solution of sodium alginate (50 mL) was added dropwise, at room temperature, to a degassed ferrous sulfate heptahydrate solution (0.3 mol L^{-1}) using a syringe fitted with a needle. The pale-yellow Fe(II)-crosslinked beads were formed immediately. The solution was gently stirred, during the whole addition, and remained under stirring for 45 min. The beads were separated by filtration and washed, several times, with ethanol/water (1/1), to remove the excess ferrous ions. Then an aqueous solution of sodium hydroxide (0.5 mol L^{-1}) was added to the beads, resulting in a change in the color of the beads from light orange to dark green. The suspension was stirred for 30 min. and then heated at 65 ± 5°C in a water bath. A solution of hydrogen peroxide (10% v/v) was added dropwise. The suspension was stirred at 65 ± 5°C for 60 min. Nitrogen was bubbled in the suspension before the oxidation stage. The resulting beads were washed with ethanol/water (1/1) and added to an aqueous solution of calcium chloride (0.3 mol L^{-1}) and stirred for 24 h. Finally, the beads were separated, washed some times and placed in a kiln at 40°C for 24 h, to obtain reddish-brown beads.

2.2 Characterization

The content of mannuronate (M) and guluronate (G) units of sodium alginate sample was 62.8% and 37.2%, respectively, calculated by 13C CP-MAS NMR in our previous study [15]. The viscosity average molar mass (¯Mv) of sodium alginate was calculated by applying the equation of Mark-Houwink-Sakurada, using K = 7.3 x 10–5 dL/g and a = 0.92, obtaining a value of 4.7 x 104 g mol^{-1} [15].

The iron oxide particle size was determined by transmission electron microscopy (TEM) using an FEI Tecnai G2 F20 microscope. The samples were ground in an agate mortar along with methyl alcohol, after which the supernatant was removed from the mortar and diluted in 1 mL of methyl alcohol. The resulting solution was kept in an ultrasound bath for 10 minutes. A drop of the mixture was then deposited on a copper mesh with carbon film. The magnetic properties were analyzed by a vibrating sample magnetometer (VSM) calibrated with a nickel standard cylinder at room temperature (25° C). The total analysis time was 10 minutes and the magnetic field ranged from 2,000 to 12,000 G, and each spot was measured at 1.5 second intervals. The Mössbauer spectrum was obtained at room temperature with constant acceleration with 1024 channels, encompassing a velocity range of −11 to +11 mm/s with increments of about 0.045 mm/s per channel. Through the X-ray diffraction (XRD) it was possible to identify the metals present in the hydrogel sample. The samples were macerated with the help of pistil and gral. The diffractograms were obtained in a Shimadzu diffractometer, model XRD

6000, equipped with iron tube and graphite monochromator. The scans were made between 15 and 60° (2θ), with a goniometer speed of 2°/ min.

The morphological properties were analyzed by granulometry using a Retsch AS 200 Basic shaker with analytical sieves of 15 to 50 mesh, optical microscopy (OM) with an Olympus SZX10 microscope, for which the samples were placed on a glass slide subjected to the action of a light beam for observation, and scanning electron microscopy (SEM) with a JOEL JSM-6510LV apparatus, after coating samples with a thin gold layer. For image capture, secondary electron detectors and energy dispersive X-ray analysis (EDX) were used, with acceleration voltage of 20 kV. The metal ion concentration in the aqueous solution and industrial wastewater was determined using a Varian AA240 atomic absorption spectrophotometer (AAS), with samples diluted in order to adjust them to the calibration curve previously obtained for each metal, followed by addition of nitric acid (HNO3) to prepare the samples for analysis.

2.3 Water uptake (WU) of nanomagnetic absorbent

The *WU* capacities were determined at room temperature through the Eq. (1).

$$WU = \frac{(W_w - W_d)}{W_d}.100 \tag{1}$$

W_w is the weight of swelled absorbent in water at equilibrium and W_d is the weight of dried absorbent (average weight of 100 mg). These experiments were performed in triplicate.

2.4 Adsorption experiments

The adsorption experiments were performed for 1 h contact time at room temperature by shaking with a constant weight of the nanomagnetic absorbent (100 mg). The initial concentrations of Mn(II) and Ni(II) in the solutions were varied from 50 mg L^{-1} and then 100 to 500 mgL^{-1} at intervals of 100 mgL^{-1}.

Adsorption experiments of the metal ions in the industrial wastewater were performed at 1 hour of contact time at room temperature by shaking with the sample natural pH (pH = 6.5). The absorbent mass ranged from 50 mg and then 100–300 mg in 100 mg intervals.

3. Results and discussion

The swelling of a polymer depends on the degree of interaction between the solvent and polymer molecules. The swelling kinetic is shown in **Figure 1**. The swelling degree at equilibrium was reached in 60 min and its value was 52%.

The size distribution curve of nanomagnetic absorbent (NA) is shown in **Figure 2(a)**. The size of the NAs produced ranged from 500 to 850 μm. The shape of the beads was mostly spherical, according to **Figure 2(b)**.

Figure 3 depicts the SEM images of NAs. In general, the beads were mostly spherical, confirming the results of OM images, and the surfaces were rough, which favors the adsorption of the NAs. The composition maps (**Figure 3**) reveal a homogeneous distribution of calcium and iron elements throughout the polymeric matrix surface. To confirm the presence of both components (calcium and iron)

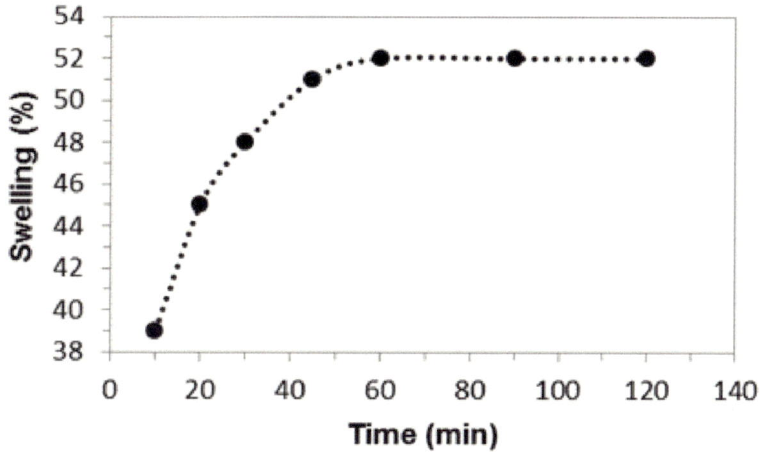

Figure 1.
Swelling kinetic of nanomagnetic absorbent, in deionized water.

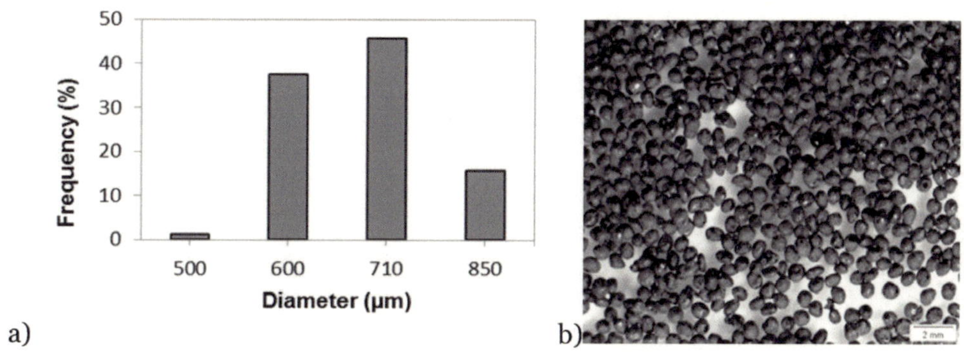

Figure 2.
Size distribution of synthesized NAs (a); optical microscopic image of NAs (b).

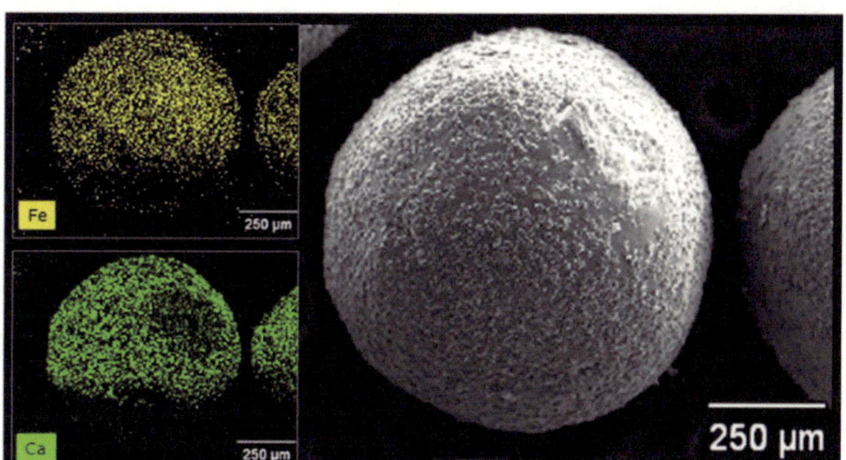

Figure 3.
SEM micrograph of NA external surface and composition maps of iron (Fe) and calcium (Ca) (magnification: 100X).

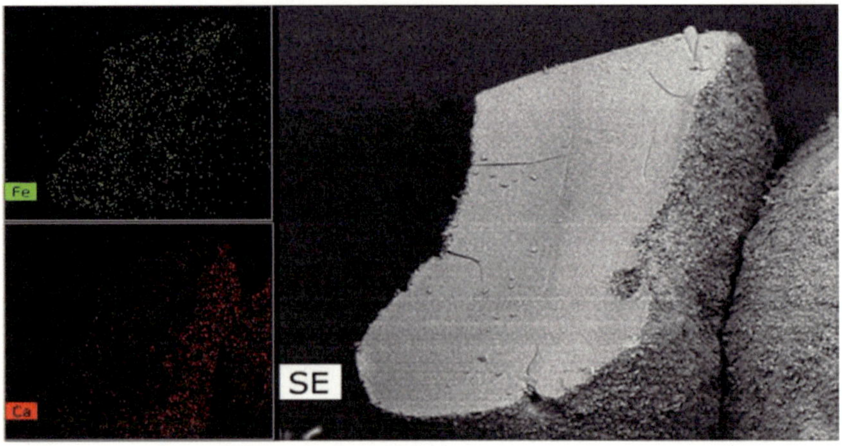

Figure 4.
SEM micrograph of NA internal surface and composition maps of iron (Fe) and calcium (Ca) (magnification: 300X).

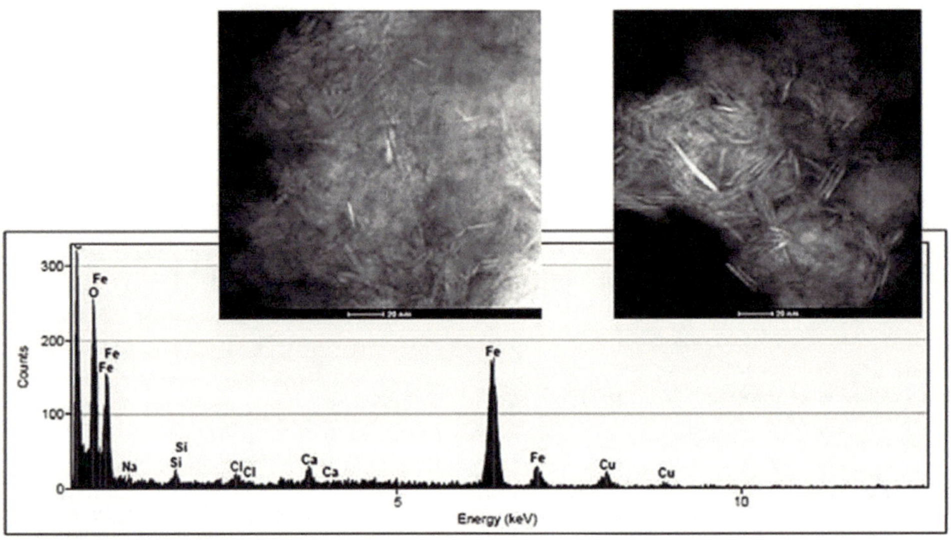

Figure 5.
TEM images of maghemite nanoparticles synthesized into the NA (in situ).

inside the microspheres, a cut as made in one of the samples ([Alg-Na] = 3% m/v; [FeSO4] = 0.3 mol L^{-1}; [CaCl2] = 0.3 mol L^{-1}) as shown in **Figure 4**. It can be seen that the microsphere has a massive internal surface containing both components.

The transmission electron microscopic (TEM) images and EDX spectra (**Figure 5**) proved that the magnetic iron oxide particles were dispersed in the NA with nanometric size (< 20 nm), confirming the formation of a nanocomposite with superparamagnetic behavior.

The VSM curve (**Figure 6**) demonstrates that the NA did not present a hysteresis cycle, a phenomenon that causes a delay between the magnetic flux density and the magnetic field. The saturation magnetization was found to be 9.88 emu/g. The values of coercivity (Hc) and remanence were 17.50 G and 0.30 emu/g, respectively. Besides that, all the samples were sensitive to magnetic stimulus by a magnet (**Figure 6**). A video is shown in Supplementar Material.

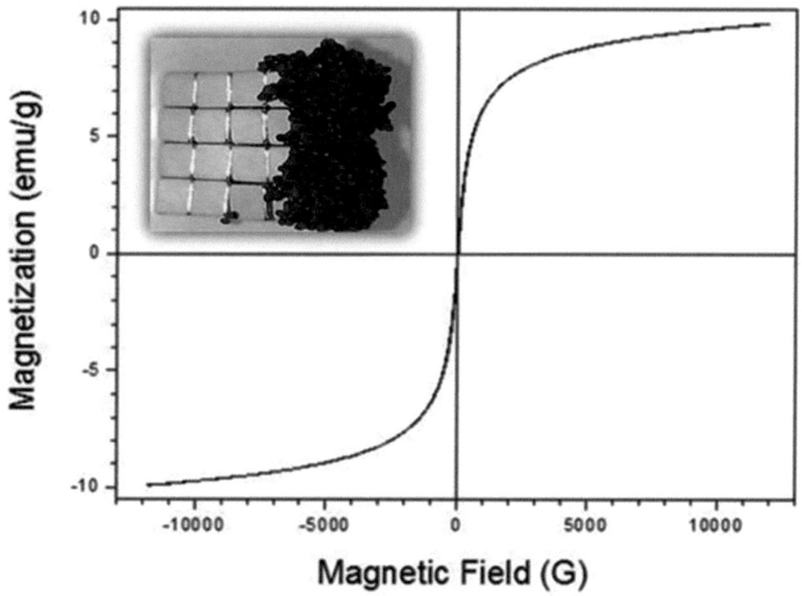

Figure 6.
Magnetization curve of NA and digital photograph of the samples responding to the stimulus of a magnet.

The Mössbauer spectrum (**Figure 7a**) showed the dominance of a central doublet and the presence of a low intensity diffuse sextet, typical of superparamagnetic relaxation [16].

The hyperfine parameters indicated the presence of a compound containing $Fe3+$ and maximum magnetic field probability less than 200 kOe. For the doublet, the quadrupole splitting and isomer shift of the sample were 0.70 mm/s and 0.35 mm/s, respectively. These values again indicate the presence of a phase containing $Fe3+$ [17, 18].

The superparamagnetic behavior is caused by the presence of nanometric particles, of the order of 10 nm. According to the XRD diffractogram (**Figure 7b**), the diffuse peaks of the carrier phase of iron indicate the presence of nanosized particles. The phase of iron (III) oxide (maghemite, γ-Fe2O3) was identified in the X-ray diffraction patterns by the rays at the approximate positions of 2θ = 36°, 45° and 52°.

3.1 Adsorption of metallic ions from industrial wastewater by the NAs

After preparing and characterizing the nanomagnetic absorbent (NA), we investigated their adsorption capacity for removal of metallic ions from wastewater samples from the metallurgical industry.

The untreated wastewater was analyzed by atomic absorption spectrometry to ascertain the concentration of metals. The results are shown in **Table 1**.

Among the metals analyzed, manganese and nickel had the highest concentrations, both derived from surface protective coating by cathodic electrodeposition, in both cases above the thresholds allowed for discharge into the environment [19]. In view of these results, we decided to evaluate the ability of the NA to adsorb these metals (Mn and Ni) to reduce their concentration.

We first performed kinetic studies of the adsorption of these metals present in an aqueous solution prepared in the laboratory to verify the adsorption equilibrium time of each metal contaminant. The results showed that the adsorption

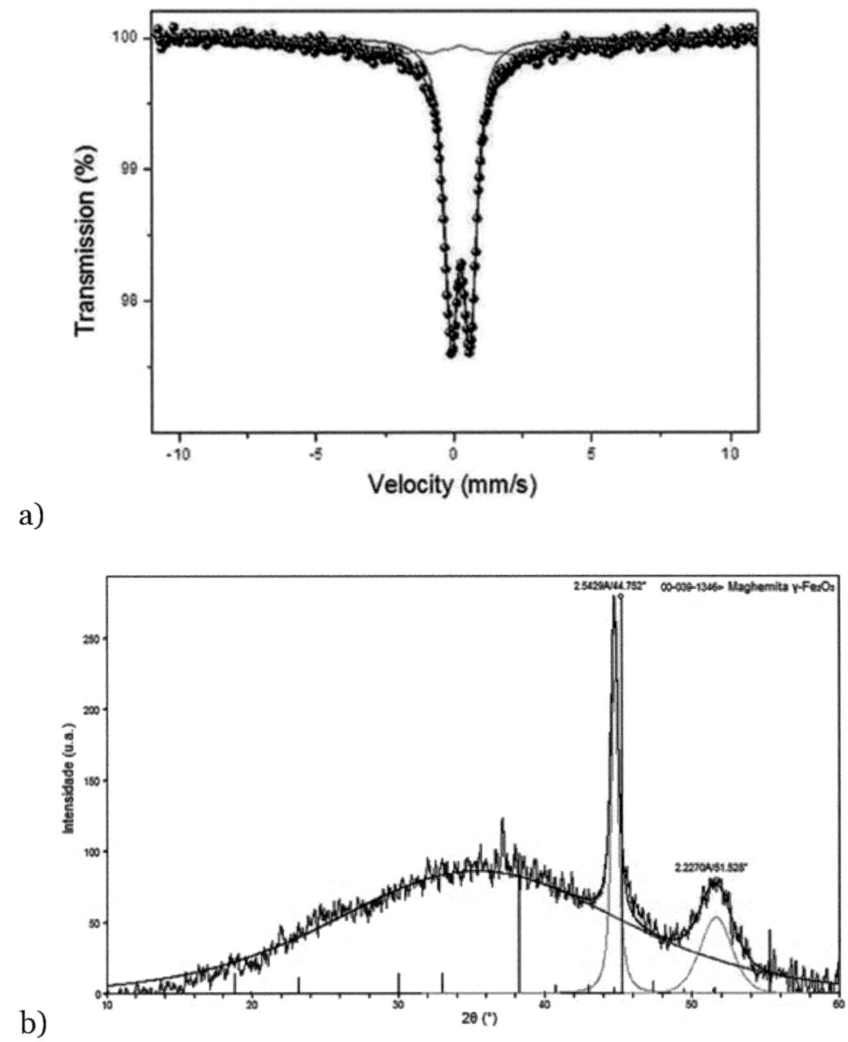

Figure 7.
Mössbauer spectrum (a) and XRD diffractogram (b) of NA.

equilibrium time for both ions was about 60 minutes, as shown in **Figure 8**, where the quantity of each metal adsorbed at equilibrium (Qe) can also be seen.

From the adsorption kinetic curves of Ni(II) and Mn(II), it can be seen that the nanomagnetic absorbent presented satisfactory removal potential. After 60 minutes, the adsorption capacity (Qe) of nickel 12.9 mg/g of NA, which corresponds to removal potential of 52%, while for manganese it was 12.3 mg/g of NA, corresponding to removal potential of 49%.

The fact that the adsorption capacity was much more pronounced in the first 30minutes can be explained by the adsorption mechanism, which occurs in three steps [20, 21]:

1. Transfer of external mass of molecules of the solute, from the body of the solution, to the surface of the adsorbent particles (transport of the adsorbate to the outside surface of the adsorbent);

2. Diffusion to the adsorption sites inside the particles' structure;

Metal	Concentration (mg L^{-1})
Al	0.20
Cd	0.06
Ca	5.46
Pb	0.07
Cu	0.06
Cr	0.00
Fe	1.38
Mn[a]	**26.88**
Ni[a]	**17.02**
Zn	1.50

[a] *Limits for discharge set by Brazil's National Environmental Council [19]: Ni = 2 mg L^{-1}; Mn = 1 mg L^{-1}.*

Table 1.
Concentration of the metals presented in the industrial wastewater sample.

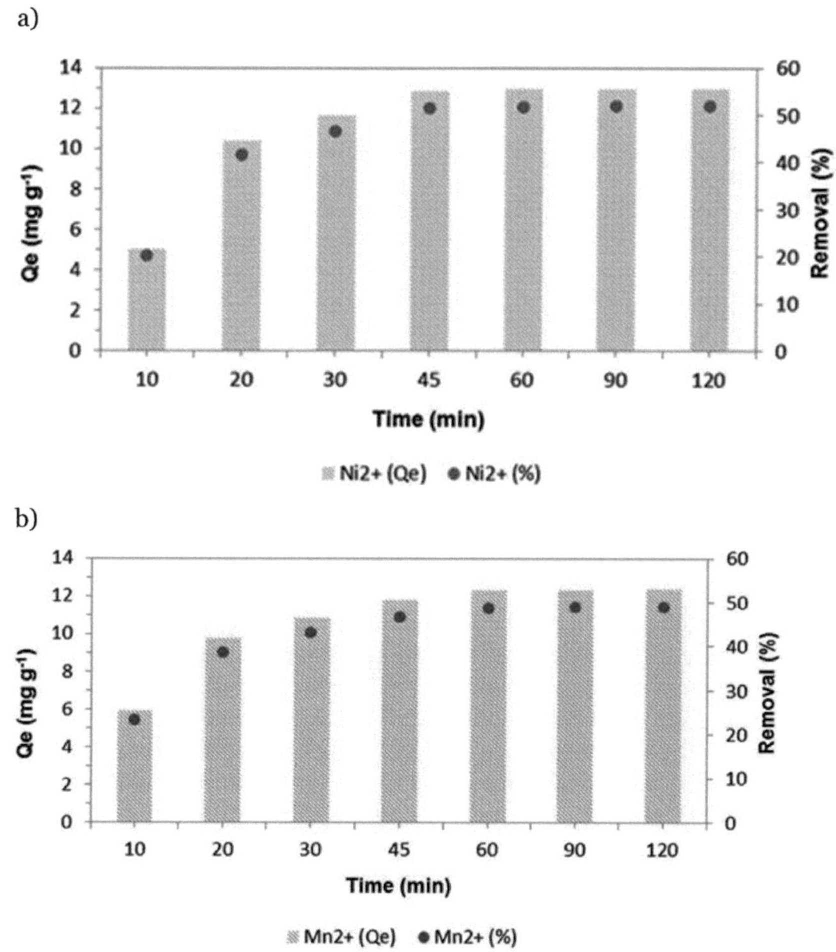

Figure 8.
Adsorption capacity (Qe) and removal percentage (%) of ions from aqueous solutions by NAs in function of time: (a) [Ni2+]onset = 50 mg L^{-1}; V = 50 mL; m = 100 mg e (b) [Mn2+]onset = 50 mg L^{-1}; V = 50 mL; mMHB = 100 mg (pH = 6.5).

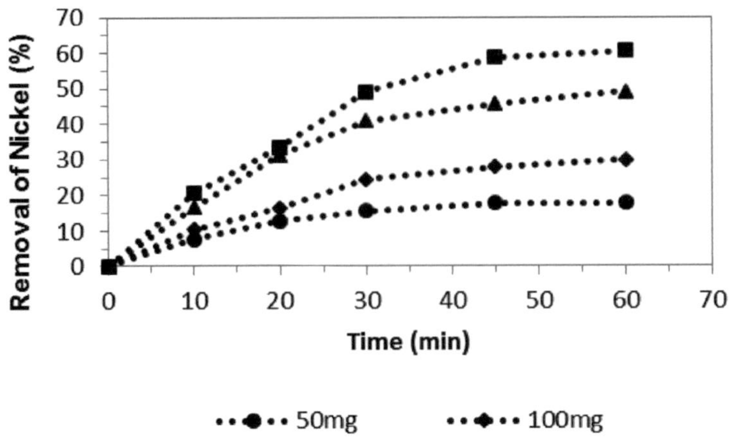

a)

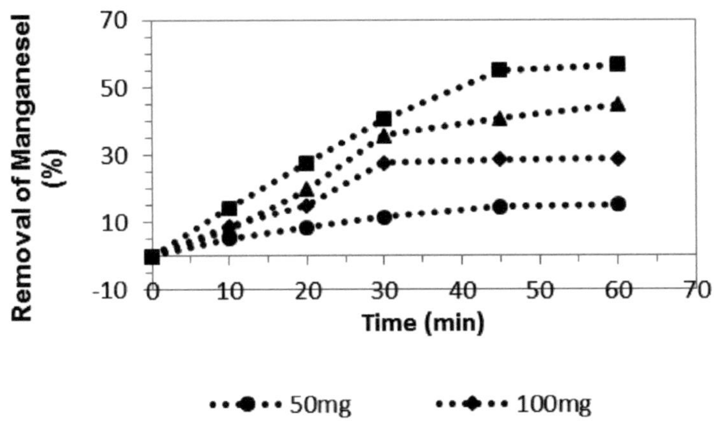

b)

Figure 9.
Effect of NA mass on adsorption of Ni(II) (a) and Mn(II) from industrial wastewater (b) [experimental condition: pH = 6.5 and V = 50 mL].

3. Immeasurably fast step, when the adsorption itself takes place (adsorption of the adsorbate on the internal surface of the adsorbent particles). The third step does not offer any resistance to the process, so that the mass transfer and intraparticle diffusion are the steps that determine the adsorption speed.

The maximum adsorption equilibrium times observed for the two metals in question were short (60 minutes), indicating rapid reaction between the adsorbent material and the metallic ions in solution. This means the NA prepared in this study has good potential for application on industrial scale.

Two experiments were performed with the industrial wastewater, one to analyze the kinetics in function of the NA mass and the other to assess the adsorption mechanism of these ions. The intention of applying the NA for commercial treatment of wastewater is to remove metallic ions without the need to adjust the pH. Therefore, all the experiments were performed at the natural pH of the sample, 6.5.

3.2 Analysis of the adsorption kinetics in function of NA mass (for treatment of wastewater)

To ascertain the most suitable quantity of nanomagnetic absorbent to remove Ni2+ and Mn2+ from the industrial wastewater, we performed studies in which the

NA mass was varied (**Figure 9**). The results showed that 300 mg of NAs removed 60.5% and 56.6% of Ni (II) and Mn (II), respectively. Also, the higher the NA mass, the greater the removal percentages of Ni (II) and Mn (II) were. That was expected, considering that with the increase of NA mass, the availability of binding sites with metals also increases.

3.3 Competitiveness between the metals: adsorption mechanism

We performed analysis by atomic absorption spectrometry of calcium, iron, nickel and manganese ions after application of the hydrogel in the wastewater sample, for the purpose of determining the interaction between these ions and the alginate (**Figure 10**). Both nickel and manganese were adsorbed by the hydrogel (concentrations in the wastewater diminished) while calcium was desorbed by

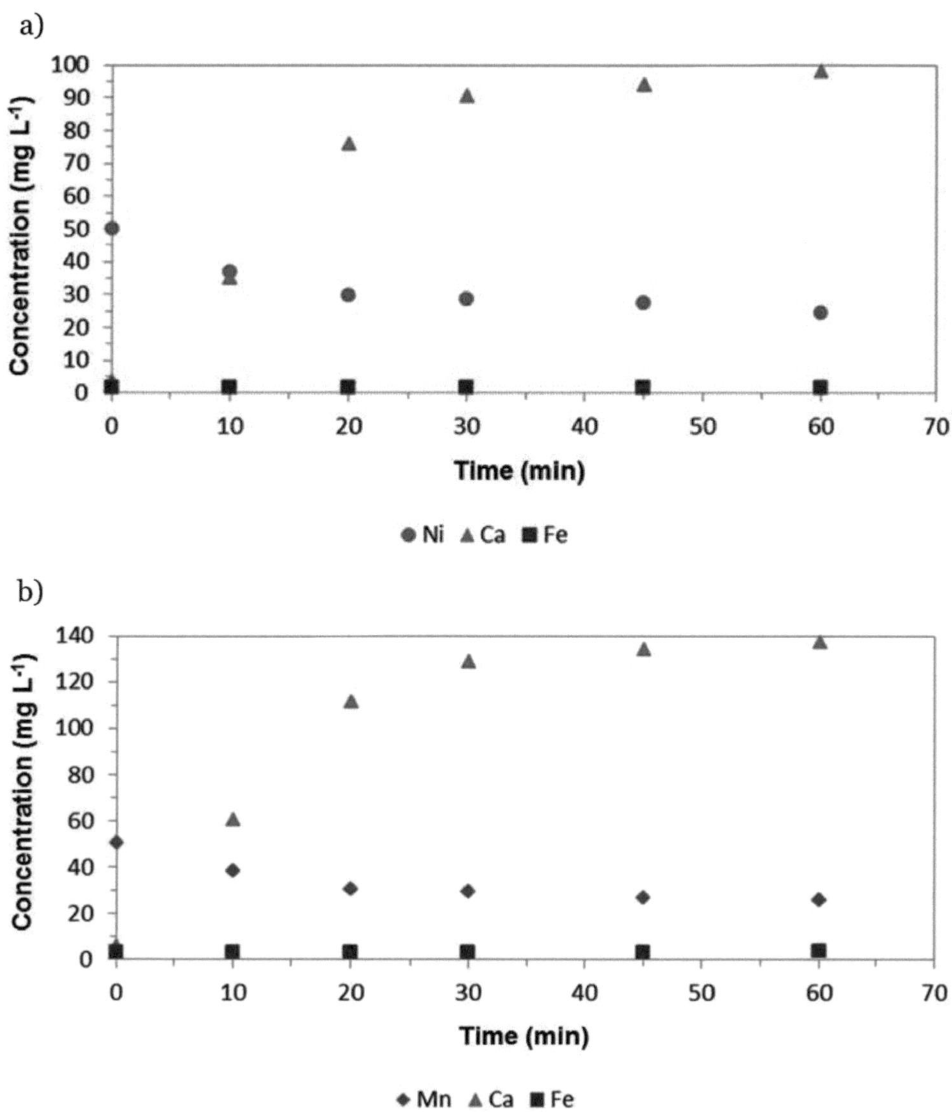

Figure 10.
Concentration of Ni2+ (a), Mn2+ (b), Ca2+ and Fe2+ ions in the wastewater as a function of contact time with the nanomagnetic absorbent (experimental conditions: mNA = 300 mg; pH = 6.5 and V = 50 mL).

the hydrogel (concentration in the medium increased). This indicates the occurrence of ionic exchange of Ni2+ and Mn2+ with Ca2+. The iron concentration remained unchanged, proving no loss of magnetic material occurred from the hydrogel.

4. Conclusion

Nanomagnetic absorbent (NA) based on calcium alginate along with gamma-maghemite nanoparticles were synthesized in situ and applied for the removal of metals present in industrial wastewater. The NA presented, predominantly, spherical morphology, with homogeneous distribution of the elements iron and calcium, both on the internal and external surfaces of the NA particles, and responded to the stimulus of a magnet. The XRD diffractogram demonstrated that the synthesized magnetic material was maghemite. The transmission electron microscopic images proved that the maghemite particles had diameters smaller than 20 nm. The NAs were good adsorbents of the metals from the wastewater. The adsorption equilibrium was achieved in 1 hour at pH of 6.5. By using 300 mg of NA, it was possible to remove simultaneously 60.5% of Ni (II) and 56.6% of Mn (II). Based on the results, we can propose that the adsorption mechanism of the metals occurred by exchange of Ca2 + ¬ ions (crosslinker) with Ni2+ and Mn2+ ions present in the wastewater.

Acknowledgements

We are thankful to the Carlos Chagas Filho Research Foundation of the State of Rio de Janeiro (FAPERJ), the National Council for Scientific and Technological Development (CNPq) for financial support. This study was financed in part by the Coordenação de Aperfeiçoamento de Pessoal de Nível Superior – Brasil (CAPES) – Finance Code 001.

Conflict of interest

The authors declare that they have no known competing financial interests or personal relationships that could have appeared to influence the work reported in this paper.

Author details

Ivana Lourenço de Mello Ferreira[1*], Rodrigo Ferreira Bittencourt[1]
and Clenilson Sousa Júnior[2]

1 State University of Rio de Janeiro, Rio de Janeiro, Brasil

2 Federal Institute of Education, Science and Technology of Rio de Janeiro,
Rio de Janeiro, Brazil

*Address all correspondence to: ivanalmellof@gmail.com

References

[1] C. Liu, R. Bai, L. Hong, Diethylenetriamine-grafted poly(glycidyl methacrylate) adsorbent for effective copper ion adsorption Journal of Colloid and Interface Science, 303, 99-108 (2006). DOI: 10.1016/j.jcis.2006.07.057

[2] Idris, A.; Ismail, N. S. M.; Hassan, N.; Misran, E.; Ngomsik, A. Synthesis of magnetic alginate beads based on maghemite nanoparticles for Pb(II) removal in aqueous solution. Journal of Industrial and Engeneering Chemistry, 18, 1582-1589 (2012). DOI: 10.1016/j.jiec.2012.02.018

[3] V. M. Nurchi, I. Villaescusa, Agricultural biomasses as sorbents of some trace metals Coordination Chemistry Reviews., 252, 1178-1188 (2008). DOI: 10.1016/j.ccr.2007.09.023

[4] Bhattacharyya K. G., Gupta S. S., Removal of Hazardous Metals from Water by Clays, in: Hazardous Materials and Wastewater: Treatment, Removal and Analysis, 79-125 (2007).

[5] L. V. A. Gurgel, R. P. Freitas, L. F. Gil, Adsorption of Cu(II), Cd(II), and Pb(II) from aqueous single metal solutions by sugarcane bagasse and mercerized sugarcane bagasse chemically modified with succinic anhydride. Carbohydrate Polymers, 74, 922-929 (2008). DOI: 10.1016/j.carbpol.2008.05.023

[6] Gupta, V.; Ali, I. Removal of lead and chromium from wastewater using bagasse fly ash – a sugar industry waste. Journal of Colloid and Interface Science, 271, 321-328 (2004). DOI: 10.1016/j.jcis.2003.11.007

[7] M. Arami, N. Y. Limaee, N.M. Mahmoodi, N.S. Tabrizi, Removal of dyes from colored textile wastewater by orange peel adsorbent: Equilibrium and kinetic studies Journal of Colloid and Interface Science., 288, 371-376 (2005). DOI: 10.1016/j.jcis.2005.03.020

[8] Paulino, A. T., Guilherme, M. R., Reis, A. V., Campese, G. M., Muniz, E. C., Nozaki, J. Removal of methylene blue dye from na aqueous media using superabsorbent hydrogel supported on modified polysaccharide. Journal of Colloid and Interface Science, 301, 55-62 (2006). DOI: 10.1016/j.jcis.2006.04.036

[9] V. K. Gupta, A. Mittal, V. Gajbe, Adsorption and desorption studies of a water soluble dye, Quinoline Yellow, using waste materials Journal of Colloid and Interface Science., 284, 89-98 (2005). DOI: 10.1016/j.jcis.2004.09.055

[10] Yu, K.; Ho, J.; Mccandlish, E.; Buckley, B.; Patel, R.; Li, Z.; Shapley, N. C. Copper ion adsorption by chitosan nanoparticles and alginate microparticles for water purification applications. Colloids and Surfaces A: Physicochemical and Engineering Aspects, 425, 31-41 (2013). DOI: 10.1016/j.colsurfa.2012.12.043

[11] Wu, D.; Zhang, L.; Wang, L.; Zhu, B.; Fan, L. Adsorption of lanthanum by magnetic alginate-chitosan gel beads. Journal Chemical Technology and Biotechnology, 86, 345-352 (2001). DOI: 10.1002/jctb.2522

[12] Bee, A.; Talbot, D.; Abramson, S.; Dupuis, V. Magnetic alginate beads for Pb(II) ions removal from wastewater. Journal of Colloid and Interface Science, 362, 486-492 (2011). DOI: 10.1016/j.jcis.2011.06.036

[13] S. F. Lim, Y. M. Zheng, S. W. Zou, J. P. Chen, Removal of copper by calcium alginate encapsulated magnetic sorbent. Chemical Engineering Journal , 152, 509-513 (2009). DOI: 10.1016/j.cej.2009.05.029

[14] Llanes, F.; Ryan, D. H.; Marchessault, R. H. Magnetic nanostructured composites using alginates of different M/G ratios as polymeric matrix. International Journal of Biological Macromolecules., 27, 35-40 (2000). DOI: 10.1016/S0141-8130(99)00115-4

[15] M. P. M. da Costa, M. C. Delpech, I. L. M. Ferreira, M. T. de M. Cruz, J. A. Castanharo, M. D. Cruz, Evaluation of single-point equations to determine intrinsic viscosity of sodium alginate and chitosan with high deacetylation degree Polymer Testing, 63, 427-433 (2017). DOI: 10.1016/j.polymertesting.2017.09.003

[16] G. M. Costa, E. Grave, R. E. Vandenberghe, Mössbauer studies of magnetite and Al-substituted maghemites Hyperfine Interactions, 117, 207-24 (1998). DOI: 10.1023/A:1012691209853

[17] G.M. Costa, C.B. Andujar, E. Grave, Q.A. Pankhurst, Magnetic Nanoparticles for in Vivo Use: A Critical Assessment of Their Composition. Journal of Physical Chemistry B, 118, 11738-11746 (2014). DOI: 10.1021/jp5055765

[18] Vanderberghe, R.E.; Barrero, C.A.; Da Costa, G.M.; Van San, E.; De Grave, E. Mössbauer Characterization of Iron Oxides and (Oxy) hydroxides: the Present State of the Art. Hyperfine Interactations 126, 247-259 (2000). DOI: 10.1023/A:1012603603203

[19] CONAMA Resolution, N° 430, 2011. Accessed on 11/22/2019. Available at: http://www.mma.gov.br/port/conama/res/res11/res43011.pdf.

[20] K. C. Justi, M. C. M. Laranjeira, A. Neves, A. S. Mangrich, V. T. Fávere, Chitosan functionalized with 2[-bis-(pyridylmethyl) aminomethyl]4-methyl-6-formyl-phenol: equilibrium and kinetics of copper (II) adsorption. Polymer, 45, 6285-6290 (2004). DOI: 10.1016/j.polymer.2004.07.009

[21] K. C. Justi, V. T. Fávere, M. C.M. Laranjeira, A. Neves, R. A. Peralta, Kinetics and equilibrium adsorption of Cu(II), Cd(II), and Ni(II) ions by chitosan functionalized with 2[-bis-(pyridylmethyl)aminomethyl]-4-methyl-6-formylphenol Journal of Colloid and Interface Science., 291, 369-374 (2005). DOI: 10.1016/j.jcis.2005.05.017

Chapter 2

Alginates - A Seaweed Product: Its Properties and Applications

S. Giridhar Reddy

Abstract

Alginates are natural polysaccharides available as seaweed products. They possess several properties due to their molecular structure made of bipolymeric α-L-Guluronic acid and β-D-Mannuronic acid polymers. Alginates have several properties such as film-forming ability, pH responsiveness, and gelling, hydrophilicity, biocompatibility, biodegradability, non-toxic, processability and ionic crosslinking. They're commonly used in several industries, including food, pharmaceuticals, dental applications, welding rods and scaffolding. Due to their gelling and non-toxic properties, as well as their abundance in nature, the cosmetics and healthcare industries have shown a great deal of interest in biodegradable polymers in general and alginates particularly over the last few decades.

Keywords: Alginates, hydrogels, biodegradable polymers, controlled drug delivery, seaweeds

1. Introduction

Polymers are the chemical compounds that play a significant role in the development of medicine and engineering with day-to-day life. Polymers are made of several repeating units of monomers, which decide the structure and properties. In the field of medicine, a range of polymeric materials is being used, however, biodegradable polymers gaining more attention.

Brown algae are composed primarily of alginates, a carbohydrate polymer found in the form of an insoluble combination of potassium, sodium, magnesium, and calcium salts of alginic acid that are structural components of brown seaweed cell walls. They are unbranched binary polymers made up of 1, 4 linkages between β-d-Mannuronic (M) and α-l-Guluronic (G) acids [1]. The composition of alginates i.e. G: M ratio varies based on the source [2]. They can also be tailored or resized in many varieties by varying molecular weight, cation content, particle form, volume fraction and, G: M ratio.

Alginates were first used in 1883 by a researcher named Edward Stanford, and commercial development commenced in 1927. The global production of alginates has now risen to about 40,000 tonnes per year. Alginates are used extensively in the food, pharmaceutical, cosmetic, and dental industries [3]. Perhaps, in recent years, the medical and pharmaceutical industries have become incredibly influential in biopolymers, specifically alginates.

Alginates are particularly known for their applications in pharmaceutical industries for controlled drug delivery [4], wound healing [5], dermatology [6],

Figure 1.
Alginate –Brown algae a seaweeds.

and scaffolds [7] because of their properties such as natural disintegration, gel formation, biocompatibility, and non-toxicity. Alginates are a natural gum that has an advantage over synthetic polymers because they form hydrogels, are less expensive, and are readily accessible. Alginate gels may also be orally administered into the body in a minimally invasive manner, enabling a wide range of pharmaceutical applications. Alginate gels are promising biomaterials for tissue engineering and cell transplantation, to replace organs in patients that have lost or failed organs or tissues (**Figure 1**) [8, 9].

2. Properties of alginates

Brown algae obtained from seaweeds such as *Laminaria hyperborea*, *Laminaria digitata*, *Laminaria japonica*, *Ascophyllum nodosum*, and *Macrocystis pyrifera* are used to make commercially available alginates [10]. In general, alginates are insoluble compounds; they are washed, crushed, dried, powdered, and treated with a basic solution especially NaOH or KOH to form sodium/potassium salt of alginic acid which is water-soluble (**Figure 2**).

Alginates are generally available in the market as Alginic acid of sodium salt or Sodium alginate [12]. Alginates due to their acidic nature make them a favorable biopolymeric biodegradable product in biomedical applications. Because of high acid content, alginates form gels due to the presence of Guluronic acid (G) monomer in alginates within a short period especially in the presence of Ca^{2+} ions. This property of gelling allows the alginate to possess multiple applications such as encapsulation of varied fragments or even cells interior of the alginate matrix with very low side effects [13]. The carboxylic group in alginates are very effective i.e. reason find many applications and can be modified based on the need [14].

Bacteria such as Pseudomonas and Azotobacter can develop bacterial alginate as an exopolysaccharide. These bacterial alginate producers may be able to produce alginates with particular monomer formulations and might be capable of producing 'tailor made' bacterial alginates using genetic and protein engineering [15].

2.1 Molecular weight

Alginates obtained from different locations in a sea bed have different molecular weights ranges between 50,000 to 5lakhs [16]. The viscosity of the alginate solution

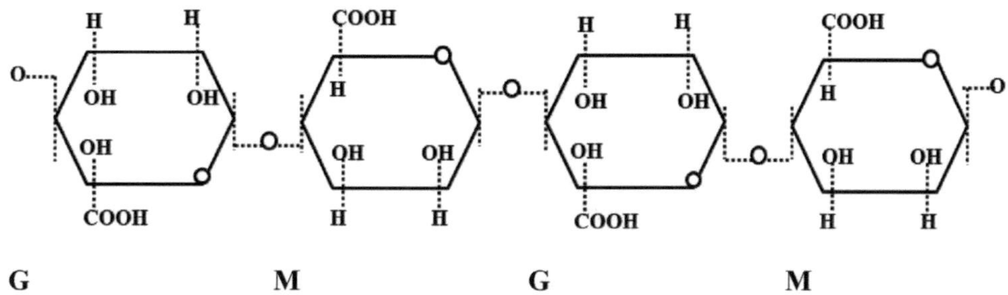

Figure 2.
Structure of Alginic acids with 1, 4 linkages between L-Guluronic acid (G) and D-Mannuronic acid (M) [11].

is pH-responsive and increase viscosity is reported with the decrease in pH and reaches a pH ~ 3.5 because carboxylate groups in Guluronic acid present in the alginate structure protonated and form hydrogen bonds [17]. Alginates may have different molecular weights depending on whether they need to monitor pre-gel solution viscosity or post-gelling strength distribution separately. To adjust the viscosity of the solution, a mixture of high and low molecular weight alginate polymers is used [18].

2.2 Biocompatibility

The biocompatibility of alginate has been extensively evaluated *in-vitro* as well as *in-vivo* at varying levels of purity. It has been reported that alginate containing high M monomers are more immunogenic and 10 times more effective in promoting cytokine synthesis compared to G monomer in alginates [19], but others reported very little response or no immune response across alginate implants [20]. Impurities in the alginates, such as heavy metals, endotoxins, proteins, and polyphenolic compounds, may be causing the variable reaction at the injection or implantation sites. However, not many serious inflammatory results were reported in commercially available or certified or alginates obtained from branded companies [21].

2.3 Alginate derivatives

Alginates are exploited or synthesized for various biomedical applications by introducing different hydrophilic moieties such as alkyl groups or hydrophilic polymers in the alginate matrix. Long-chain alkyl groups such as dodecyl or octadecyl are bonded with alginates matrix *via* esterification. The various properties such as rheology, gelling and crosslinking characteristics are very useful in bone regeneration and cartilage repair [22]. Sustained or controlled drug delivery vehicles are achieved from alginate derived from Poly (butyl methacrylate) [23].

Alginates are also investigated for their derivatives with cell-adhesive peptides are prepared by adding peptides as side chains and coupled through the carboxylic groups of the sugar residues [24, 25].

2.4 Gelling properties

The ability of aqueous alginate solutions to form gels when treated with divalent ions (Ca^{2+}, Sr^{2+}, and Ba^{2+}) or trivalent ions (Fe^{3+} and Al^{3+}) ions has been extensively explored for the fabrication of carriers for sustained or controlled delivery of therapeutic agents. This is due to intramolecular bonding and ionic interactions that exist within the carboxylic acid groups on the polymer matrix and the cations present as shown in **Figure 2** [26, 27].

The calcium or any other divalent or trivalent ions will interact with the G monomer present in the alginate structure to crosslink with another molecule, and the structure is identical to the egg box model (**Figure 3**) [28].

The complexing forming agents such as EDTA-sodium citrate [29] or monovalent cations, complex anions (phosphate, and citrate) which have a high affinity for Ca^{2+}ions, which can disrupt calcium alginate gels easily. The presence of high concentrations of non-gelling ions (Na^{+} and Mg^{2+}) also contributes to the instability. It was reported that the strength and uniformity of gelling largely depend on the type of crosslinking and temperature [30].

It is also reported that the strength of alginate film with Al^{3+} ion is very low compared to other divalent ions (Ca^{2+} and Ba^{2+}) crosslinking because in the case of Al^{3+} ions the crosslinking occurs in two different planes of the alginate structure and at the same time it makes the alginate framework more compact [31]. The small size of the Al^{3+} ion (0.58 A) facilitates its diffusion into the matrix of the film without crosslinking on the surface and thus results in poor crosslinking [32].

For a variety of uses, including tissue engineering, alginates are being investigated for covalent bonding to enhance the physical properties of gels. In the case of alginates with covalent bonding degradation of chain occurs even with a slight increase in temperature due to breaking of crosslinks ultimately leading to stress relaxation due to water migration. It was reported that covalently bonded alginates are found to be toxic [33]. Furthermore, the Ca^{2+} ions released out of the gel can boost hemostasis, while the gel acts as a matrix for platelet and erythrocyte aggregation [34].

To prepare gels with a wide spectrum of mechanical properties, covalent crosslinking of alginate with Poly (ethylene glycol)-diamines of different molecular weights was first investigated. The elastic modulus showed improvement in gel with crosslinking density or weight fraction of Polyethylene glycol (PEG), but then decreased as the molecular weight between cross-links increased became less compared to the original PEG [35]. Using various types of cross-linking molecules and regulating cross-linking densities, it was later investigated that the mechanical properties, as well as swelling of alginate strength, are tightly regulated. As most would imagine, the chemistry of the cross-linking molecules has a major impact on hydrogel swelling. While shaping hydrogels, multi-functional

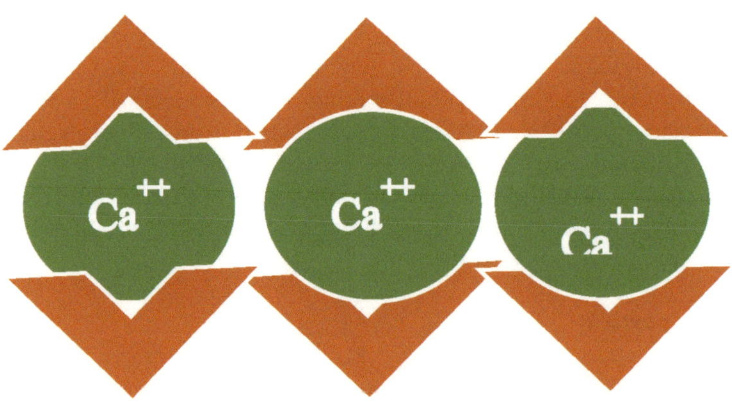

Figure 3.
The divalent calcium cation binds in the cavities of alginates like eggs in an egg box.

cross-linking entities have a greater scope of degradation efficiency and mechanical strength performance than bi-functional cross-linking molecules. Physical properties and degradation behavior of Poly-aldehyde guluronate (PAG) gels are prepared with either Poly-acrylamide-co-hydrazide (PAH) or Adipic acid dihydrazide (AAD) as a cross-linking agent were investigated *in-vitro*. PAG/PAH gels had higher mechanical stiffness and very low degradation is observed compared to PAG/AAD gels [36].

Photo crosslinking is used for crosslinking in alginates structures under mild conditions either direct or indirect sunlight or in the presence of suitable initiators. For example, the mechanical properties of Polyallylamine and alginate were significantly improved from this technique [37].

2.5 Solubility

The alginates with divalent or trivalent cations are not soluble in water because the alginates contain a terminal carboxylic ion (–COO-), so these cations bond to this and yield an insoluble product. As a consequence, the alternative is the absorption of as much as 200–300 times their weight in water, thus swelling to a hydrogel of paste-like consistency. However, alginates with monovalent cations (Na^+, K^+, and $NH4^+$) are soluble in hot and cold water. Alginates have a wide range of solubilities due to their different molecular weights.

Alginates derived from Ascophyllum, for example, have aqueous solubility in the range of 22–30% weight percent, whereas those of two Laminaria groups are 17–33% weight percent and 25–44% weight percent, respectively [38, 39].

2.6 Effect of pH on alginates

The solubility of alginates was influenced by parameters such as pH and ionic strength. Alginates have very low solubility in lower pH values due to the deprotonation of carboxylic groups (–COO-). The viscosity of alginates is unaffected above pH > 5, whereas solution having pH < 5, the COO- group present in alginates gets protonated to –COOH, and the electrostatic repulsion between chains decreases, they can move closer together to form hydrogen bonds, whereby the viscosity decreases [40]. It has been reported pH > 11 alginates viscosity is reduced due to de-polymerization [41].

The concentration of ionic solution influences the crosslinking, which increases the viscosity and molecular weight of alginates [39, 42]. Furthermore, the crosslinking depends on the confirmation of monomers G and M groups present in the alginate matrix.

2.7 Sterilization

It is reported that the viscosity of alginates decreases with autoclave sterilization because the heating randomly breaks alginate chains. The degree to which this loss occurs is determined by the presence of other kinds of stuff in the solution. Alginate solutions have also been sterilized using γ-radiation and ethylene oxide [43].

2.8 Immunogenicity

Control drug delivery is the latest trend in the presence of pharmaceutical dosage requirements for successful application in drug carriers; alginates play an important role because of their biocompatibility and immunogenicity [44].

The two key factors responsible for alginate immunogenicity are its chemical composition and the mitogenic pollutants present in alginates [45]. When alginate comes into contact with blood, it is believed to have mild cytotoxic effects and decreased hemolysis.

2.9 Biocompatibility

The biocompatibility and strength of alginic acid are determined by the quantity and quality of the acid. Furthermore, the impacts of the quantity of G monomer on alginate biocompatibility are still under investigation. Experts find different opinions on the G content of extremely purified alginate rich in G monomer residues, while others have emphasized the importance of high purity while ignoring the effects of chemical composition [46]. Animals such as rats are injected in their kidney parts with calcium alginate for biocompatibility studies and the results obtained are very promising [45].

Alginates are also experimented on in mammals and observed that they could not digest because they lack the enzyme '*Alginase*' that can break the polymer chains. Whereas ionically cross-linked alginate gels undergo dissolution by replacing the divalent crosslinked gel with monovalent cations into the surrounding media. Even alginates dissolve in the body they cannot be expelled from the body because of its high molecular weight are higher than renal clearance [47]. It is reported, alginates were obtained from *Undaria pinnatifida*, a brown seaweed invasive in the Argentinian coast are found to be toxic, but its purification using commercial techniques improves its biocompatibility and eliminates cytotoxicity in an alginate matrix for bone tissue engineering [48].

2.10 Bioadhesion

The binding or contact between two surfaces, one among being a biological substrate, is known as bioadhesion [49]. Mucoadhesion is an example in which the mucosal layer used. The carboxyl group in alginates represents a mucoadhesive anionic polymeric layer. It was reported polyanion polymers are more efficient bioadhesives compared to polycation or non-ionic polymers [50]. Alginate has better mucoadhesive strength as compared to polymers like Polystyrene, Chitosan, Carboxymethyl cellulose, and Poly (lactic acid). The bioadhesive properties of alginate would be advantageous as a mucosal drug delivery vehicle to the GI tract and nasopharynx by extending drug residence time at the site of action making them more effective [44, 51, 52].

2.11 Toxicity

Plenty of studies are reported that alginates especially crosslinked sodium/calcium alginates are non-toxic to cells, even not shown any irritation to eyes and skin [53]. Because of the nontoxicity, they found various applications in drug delivery, cosmetics, and food industries.

3. Applications of Alginates

Alginates are available in plenty in oceans and because of their diverse properties such as biodegradation, biodegradable, non-toxic, etc., as mentioned above; they have plenty of applications in the food industry, pharmaceuticals, cosmetics, textile industry, welding, and animal feeds, etc.

3.1 Food industry

FAO/WHO approves that alginates are the safest food additives because of its unique properties in food applications [54]. Over decades, alginates are used in a variety of food items, such as ice cream toppings, fruit jams, jelly, milk products, food packing, instant noodles, beer, etc., because of their unique properties such as thickening, gelling, emulsification, stabilization, texture functionality.

3.2 Pharmaceuticals

In the present medical field, many medicines or drugs administered to patients are causing a lot of side effects. Hence, there's a lot of demand created in the world for drug loading carriers which increase drug resident time *in-vitro* or *in-vivo*, especially in the gastrointestinal tract and on the exterior part of the human body should be safe and non-toxic. The developments of alginates and alginate derivatives are discussed about controlled drug delivery, sustained drug delivery, and targeted drug delivery, etc.

3.2.1 Controlled drug delivery

In general, the structure of alginates gels shows it possesses porous size (~5 nm) helps to fill this gap with small molecular weight drugs through either physical or chemical bonding. When a drug-loaded/embedded drug comes in contact with an aqueous medium, the drug release is controlled. Furthermore, the drug-loaded carriers are water-soluble and may undergo degradation in an aqueous medium, hence crosslinking the alginates with bivalent or trivalent cation will enhance the stability of the gels or films. These properties help us to study the kinetics of drug release. The Sodium salt of alginic acid (SA) and Polyethylene oxide blends are investigated for controlled release of Valganciclovir hydrochloride *in-vitro*, as an anti-HIV drug [55]. Floating microbeads made of SA and modified Chinese yam starch investigated for controlled delivery of Metformin hydrochloride drug [56]. SA and Chitosan blend with different wt% of Montmorillonite (Cloisite 30B) solution studied for controlled release [57]. An anti-cancer drug such as Paclitaxel was loaded and investigated for release *in-vitro* in variable pH medium, time, and drug concentrations. SA and Xanthum gum blends crosslinked with zinc acetate loading Ranitidine hydrochloric acid drug *in vitro* release investigated in simulated intestinal fluid (pH 7.5) and simulated gastric fluid (pH 1.2) [58]. Oral delivery of protein drug is explored for Bovine serum albumin (BSA) using composite microparticles made of Chitosan, SA, and Pectin crosslinked by tripolyphosphate [59]. A pH-responsive Tamarind seed polysaccharide and alginate blends are studied for controlled release of Diclofenac sodium. Swelling and degradation studies were also investigated in different pH mediums [60].

An anti-inflammatory drug such as Nifedipine is investigated for microspheres made of SA and Methylcellulose using Glutaraldehyde as crosslinking agent [61].

The Acrylamide and Poly(vinyl alcohol) beads with SA are grafted on exposure to UV radiation and further crosslinked with glutaraldehyde. The crosslinked beads are used to study the controlled release of Diclofenac sodium drug [62].

The transdermal films are synthesized using SA and Xanthum gum. The films loaded with Ketoprofen drug is studied *in vitro* for skin permeation [63].

SA mixture with Sodium Carboxymethylcellulose, Carbopol-934, and Polyvinyl pyrrolidine and backing membrane (Ethylcellulose), and Glycerol as plasticizer give promising results for control of drugs [64].

Hypertension drugs such as Felodipine were investigated for control release from alginate microspheres in combination with a mixture of Hydroxypropyl methylcellulose, Eudragit RS 30D, and Chitosan in simulated intestinal media [65].

Bioadhesive ocular insert of Ciprofloxacin hydrochloride using SA as gel and Chitosan as bioadhesive agent, Glycerin as a plasticizing agent are used *in-vitro* release studies were carried [66].

3.2.2 Protein delivery

The proteins drugs are in high demand and thanks to advances in recombinant DNA technology, a diverse variety of protein drugs are now available. Alginates are an ideal candidate for protein drug delivery because they can load into the alginate matrix at different formulations at relatively mild conditions which can avoid denaturation as well as degradation. A variety of techniques are explored for the controlled release of protein from alginate gels. Alginates are known for minute pores due to the presence of G and M blocks because of their bi-polymeric structural arrangements. The vascular endothelial growth factor binding to alginate hydrogels is successful in sustained and localized release [67, 68]. It is reported that alginate microspheres are efficiently loaded with lysozyme and chymotrypsin for sustained release [69]. Oral delivery of proteins such as Amino group-terminated Poly((2-dimethylamino) ethyl methacrylate with alginate gel beads was prepared [70]. Alginates are explored for stimuli-responsive gels in the synthesis of tetra-functional acetal-linked polymer networks with adjustable pore sizes.

Bovine serum albumin is used as a model protein using blends made of alginate, chitosan, and pectin composite mixture [71]. In another work, hemoglobin as a model protein loaded in poly (L-histidine)-chitosan/alginate microcapsules are reported [72, 73].

3.2.3 Wound dressing

Any injury, burns, torn, muscle pains, cuts occurring on the human body may take a longer duration for curing for even applying any ointments such as antiseptic, antimicrobial, anti-inflammatory, antipruritic, pain-relieving gels, anti-mycotic with fungal action may cause skin irritation or side effects. Hence alginates due to their appreciable properties and due to their non-toxic are extensively used in wound healing to load appropriate drugs in alginate gels and which increases the retention time of the drug, so that the drug release in small dosages on the specific site. Alginates also have hemostatic properties, making them useful in the treatment of bleeding wounds. Several investigations are reported which initially experimented on animals such as mice, pigs, and rabbits. Alginates are employed to make hydrogels, films, wafers, foams, nanofibers, including therapeutic formulations for wound dressings. Alginate wound dressings readily absorb wound fluid, maintain a physiologically moist atmosphere, and protect against bacterial infections at the wound site. Since alginates are poor in mechanical strength are blended with other polymers or composites to improve film strength. The amount of M-block presents in alginate influences the immunogenic effect and M-block induces cytokine production (**Table 1**; **Figure 4**) [81].

3.3 Alginates in cosmetics

Scientists have been researching alginates for decades to create high-quality cosmetic products that provide all benefits to the skin. Alginates, being marine

Applications	Alginate based dressings for wound healing	Refs.
Ulcer	1. Algicell™ 2. AlgiSite M™ 3. Comfeel Plus™ 4. Kaltostat™ Sorbsan™ 5. SeaSorb	[74–77]
Diabetic wounds	1. Tegagen™ 2. Hyalogran®	[78]
Thyroid and spine surgeries.	1. Guardix-^{SG}	[81]
Necrotic wounds	1. Algivon	[79]
Burns	1. Fibracol™ Plus	[80]

Table 1.
Some of the commercially available alginate-based dressings.

Figure 4.
Application of Alginate in wound dressings [82].

plants known to absorb UV rays, repair the sun's harmful effects, moisturize the epidermis, skin smoothening, and ensure the small cells renewal [83]. Alginates, in general, are known for gel formation can thicken and maintain moisture, are used in a variety of cosmetics. By forming a gel network, alginate assists in the retention of lipstick color on the lips and face creams as well as body lotion moisturizers. To make stable all-around lotions, alginate a natural thickener is incorporated in sunflower wax. Polysorbate-20 is the best commercial product is manufactured which is a smooth lotion in which instant cold emission products are combined with an emulsifier [84]. Alginates are a natural polysaccharide that possesses a very high viscosity and has a strong potential for water absorption. Perhaps, alginate's viscosity can be optimized to ensure maximum viscosity. Alginates are explored for anti-aging masks and face masks which slow down the aging process, even wrinkles and lift the skin. Also, alginates are explored in Dentures which removable set of replacement teeth and gum tissue that can provide you with the complete backing and attractive appearance you need, even as you age.

Alginates application indentures are a removable set of restorative dentistry and gum tissues that can provide us with both the complete backing and beautiful appearance even at older age (**Figures 5** and **6**) [87].

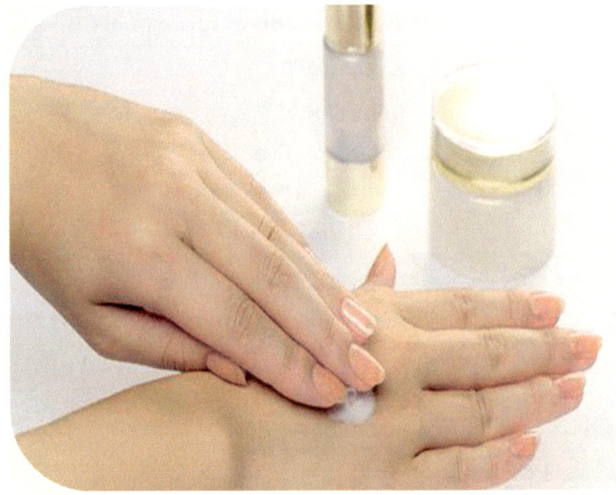

Figure 5.
Cosmetics made of alginates [85].

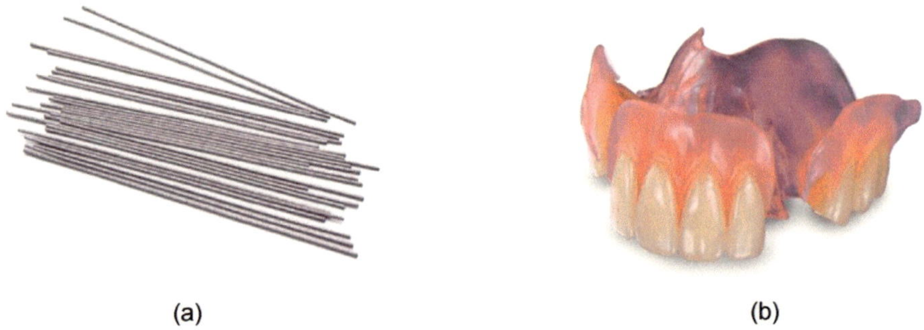

Figure 6.
Application of alginates in a) dental solders [85] and b) flexible dentures [86].

3.4 Alginates in the textile industry

Color paste substrates are prepared using textile-grade alginates in the application of patterns in print fabrics, shawls, towels, and other products. Alginates are cleaner and easier to decompose substrate for textile printing than other substrates. Alginates application in printing cotton, jute, and rayon allows for easier disposal of wastewater. Sodium alginates are thickeners used in textile printing to thicken the dye paste.

Screen or roller printing devices may be used to apply the pastes to the cloth. Alginates became common thickeners with the discovery of reactive dyes. The cellulose in the cloth reacts chemically with these substances. Many popular thickeners, such as starch, react with reactive dyes, resulting in lower color yields and sometimes difficult-to-wash-out by-products. Alginates are the strongest thickeners for reactive dyes because they non-reactive with the dyes and quickly washout from the finished textile.

In older screen printing the alginate of medium to high viscosity is used, whereas in modern high-speed roller printers even low viscosity alginates are giving very attractive printing.

3.5 Welding rods

Welding is the technique finds application building all kind of structures with metals. In the welding, process coating is used as a flux and to monitor conditions near the weld, such as temperature, and oxygen, and hydrogen. In this case, sodium silicate (water glass) is mixed with the dry coating ingredients to provide some of the plasticity needed for coating extrusion into the rod and to tie the dried coating to the rod. The silicate, on the other hand, neither binds nor provides enough lubrication to allow for successful and smooth extrusion. A lubricant and a binder are needed to keep the damp mass together before extrusion and to keep the coating on the rod in form throughout drying and baking [88]. To achieve these standards, alginates are used.

3.6 Alginates used in animal feeds

Alginate salts of sodium and potassium are meant to be used as industrial substances as emulsifiers, stabilizers, thickeners, gelling agents, and binders. There is no competent authority that has recommended the usage of sodium alginate in feeding stuffs for dogs, other non-food-producing animals, and fish. Whereas potassium alginate is used in cat and dog food at a speck of 40 g/kg feed [89]. The use of alginates in fish feed has no harmful effect on the consumer. Alginates are said to be slightly irritating to the eyes but not to the skin. The use of these ingredients in fish feed poses no threat to the aquatic environment.

A gel-type livestock feed mixture is formed by mixing feed nutrients, water, alginate, and a water-insoluble calcium component to resist the calcium content from reacting with the alginate. The calcium component is solubilized or the sequestrate affecting the reactivity between the alginate and the calcium component is extracted after the feed mixture is formed, resulting in a gel feed containing a gel matrix containing the feed nutrient ingredients. The livestock can then be fed the gel meal [90].

4. Conclusions

Alginate is a seaweed product which is a polysaccharide or carbohydrate polymer has plenty of applications in day-to-day life due to its potential characters. The Alginate properties such as biocompatibility, biodegradability, hydrogel formations, nontoxic, pH-responsive, and its ability to form a gel with crosslinking agents find their application in controlled, sustained, and targeted drug delivery carriers for a wide range of drugs or any other bioactive agents. Alginates films like any other hydrogels lack physical strength, which can be overcome by crosslinking with different divalent or trivalent cations. A continuous challenge was on researchers to explore the possibility of modification of alginate structure by grafting with other low or high molecular weight polymers to improve its strength based on the need of the application. Biomedical applications, specifically in wound healing *in -vitro* cell culture, and tissue engineering, look interesting as a potential biomaterial. Acid-responsive drugs and drugs that irritate the gastric mucosa can also be released with alginate. Alginates in combination with other polymer additives or nanoparticles are used in wound healing in diabetic wounds, ulcers, and burns. Alginates are used in an increasing number of food-related applications and have the potential to be used in a growing number of food-related applications. The gelling property of alginates makes them ideal for use as thickeners, stabilizers. Alginate consumption has several physiological effects in the gastrointestinal tract and in the body that are similar to a variety of other viscous polysaccharides. Dentures made of alginates composites are found to be very flexible, removable, and compact with the gum tissues. The high viscosity of alginates finds suitable thickeners in the textile industry as a paste containing a dye that can be applied to the fabric by either screen or roller printing. Potassium alginates and gel-type animal feeds are found to highly nutritious for cats, fishes, and piglets.

Author details

S. Giridhar Reddy
Department of Chemistry, Amrita School of Engineering, Amrita Vishwa Vidyapeetham, Bengaluru, India

*Address all correspondence to: s_giri@blr.amrita.edu

References

[1] Ott C.M, Day D.F., Marcel Dekker, J.J. Inc. New York, Chap. 5.2000.

[2] Domb AJ, Wiseman DM. Handbook of Biodegradable Polymers. Boca Raton, CRC Press.1998.

[3] Ertesvag H, Valla S. (1998): Biosynthesis and applications of alginates. Polymer Degradation and Stability, 5, 985.

[4] Hanne Hjorth Tønnesen , Jan Karlsen. Drug Development and Industrial Pharmacy , Volume 28, 2002 - Issue 6, doi.org/10.1081/ DDC-120003853

[5] Alginate dressings, Hand book of medical textiles. 2011

[6] Maedeh Bahadoran, Amir Shamloo, Yeganeh Dorri Nokoorani , Scientific Reports . 2020; 10: 7342.

[7] Maryam Farokhi, Farinaz Jonidi Shariatzadeh, Atefeh Solouk , Hamid Mirzadeh,International Journal of Polymeric Materials and Polymeric Biomaterials. Alginate Based Scaffolds for Cartilage Tissue Engineering: A Review, 2020 69(4): 230-247. doi.org/10.1080/00914037.2018.1562924.

[8] Langer R, Vacanti JP. Tissue engineering. Science. 1993;260 :920-926.

[9] Lee KY, Mooney DJ. Hydrogels for tissue engineering. Chem Rev. 2001;101:1869-1879

[10] Smidsrod O, Skjak-Bræk G. Alginate as immobilization matrix for cells. Trend Biotechnol. 1990;8:71-78. doi.org/10.1016/0167-7799(90)90139-O.

[11] Soares PJ, Santos JE, Chierice GO, Cavalheiro ETG. Thermal behavior of alginic acid and its sodium salt, Eclet. Química. 2004, .29 (2): 57-64. Doi.org/10.1590/S0100-46702004000200009.

[12] Kuen Yong Lee, David J. Mooney. Alginate: Properties and biomedical applications. Prog Polym Sci. 2012, 37(1): 106-126. doi:10.1016/j.progpolymsci.2011.06.003.

[13] Klock G, Pfeffermann A, Ryser C, Grohn P, Kuttler B, Hahn HJ, et al. Biocompatibility of mannuronic acid-rich alginates. Biomaterials. 1997, 18:707-713. DOI: 10.1016/ s0142-9612(96)00204-9

[14] Luca Szabó, Sandrine Gerber-Lemaire,* Christine Wandrey, Strategies to functionalize the anionic biopolymer Na-Alginate without restricting its polyelectrolyte properties, Polymers (Basel). 2020; 12(4): 919. doi: 10.3390/ polym12040919.

[15] Lain Hay D, Zahid Ur Rehman , Aamir Ghafoor , Bernd H. A. Rehm, Bacterial biosynthesis of alginates, Journal of chemical tech and Biotech. 2010;85(6): 752-759. doi.org/10.1002/ jctb.2372.

[16] Lakshmi S. Nair, Cato T. Laurencin. Biodegradable polymers as biomaterials. Prog. Polym. Sci. 32 (2007) 762-798. doi:10.1016/j.progpolymsci.2007.05.017.

[17] Rinaudo M. On the abnormal exponents' aη and aD in mark-Houwink type equations for wormlike chain polysaccharides. Polym Bull. 1992; 27:585-589. doi.org/10.1007/ BF00300608.

[18] Kong HJ, Lee KY, Mooney DJ. Decoupling the dependence of rheological/mechanical properties of hydrogels from solids concentration. Polymer. 2002; 43:6239-6246. doi.org/10.1016/S0032-3861(02)00559-1.

[19] Otterlei M, Ostgaard K, Skjakbraek G, Smidsrod O, Soonshiong P, Espevik T. Induction of cytokine production from human monocytes stimulated with alginate. J Immunother, 1991; 10: 286-291. doi:10.1016/j.progpolymsci.2011.06.003.

[20] Zimmermann U, Klock G, Federlin K, Haning K, Kowaslski M, Bretzel RG, Horcher A, Entenmann H, Siebers U, Zekorn T. Production of mitogen contamination free alginates with variable ratios of mannuronic to guluronic acid by free flow electrophoresis. Electrophoresis. 1992;13:269-74. doi.org/10.1002/elps.1150130156.

[21] Lee J, Lee KY. Local and sustained vascular endothelial growth factor delivery for angiogenesis using an injectable system. Pharm Res 2009; 26:1739-1744. doi: 10.1007/s11095-009-9884-4

[22] Pelletier S, Hubert P, Payan E, Marchal P, Choplin L, Dellacherie E. Amphiphilic derivatives of sodium alginate and hyaluronate for cartilage repair: Rheological properties. J Biomed Mater Res. 2001; 54:102-108. DOI: 10.1002/1097-4636(200101)54:1<102::aid-jbm12>3.0.co;2-1.

[23] Yao BL, Ni CH, Xiong C, Zhu CP, Huang B. Hydrophobic modification of sodium alginate and its application in drug controlled release. Bioprocess Biosyst Eng. 2010; 33:457-463. doi: 10.1007/s00449-009-0349-2.

[24] Lehenkari PP, Horton MA. Single integrin molecule adhesion forces in intact cells measured by atomic force microscopy. Biochem Biophys Res Commun. 1999; 259:645-650. https://doi.org/10.1006/bbrc.1999.0827.

[25] Koo LY, Irvine DJ, Mayes AM, Lauffenburger DA, Griffith LG. Co-regulation of cell adhesion by nanoscale RGD organization and mechanical stimulus. J Cell Sci. 2002; 115:1423-1433. doi.org/10.1242/jcs.115.7.1423

[26] Miyazaki S, Kubo W, Attwood D. Oral sustained delivery of theophylline using in-situ gelation of sodium alginate. J. Control Release 2000, 67 (2-3), 275-280. doi.org/10.3109/03639045.2011.634809.

[27] Katchalsky A, Cooper RE, Upadhyay J, Wassermann A. Counter-ion fixation in alginates. J. Chem. Soc. 1961; 5198. doi.org/10.1039/JR9610005198

[28] Grant GT, Morris ER, Rees DA, Smith PJC, Thom D. Biological interactions between polysaccharides and divalent cations: The egg-box model. FEBS Letters. 1973; 32 (1):195-198. doi.org/10.1016/0014-5793(73)80770-7.

[29] Smeds KA, Grinstaff M.W. Photocrosslinkable polysaccharides for in situ hydrogel formation. J. Biomed. Mat. Res. 2001, 54 (1),115-12 1. doi.org/10.1002/1097-4636(200105)55:2<254::AID-JBM1012>3.0.CO;2-5.

[30] Kuo CK, Ma PX. Ionically crosslinked alginate hydrogels as scaffolds for tissue engineering: Part structure, gelation rate and mechanical properties. Biomaterials. 2001; 22: 511-521. doi.org/10.1016/S0142-9612(00)00201-5.

[31] S. Al-Musa, D. A. Fara, and A. A. Badwan. Evaluation of parameters involved in preparation and release of drug loaded in crosslinked matrices of alginate J. Controlled Release, 57, 223 (1999). doi: 10.1016/s0168-3659(98)00096-0.

[32] S. Giridhar Reddy, Akanksha Saxena Pandit, and Amrita Thakur. Study on

the Effects of crosslink agents on Sodium Alginate and Lignosulphonic Acid Blends Polymer(Korea), Vol. 40, No. 1, pp. 63-69 (2016). DOI:10.7317/pk.2016.40.1.63.

[33] Zhao XH, Huebsch N, Mooney DJ, Suo ZG. Stress-relaxation behavior in gels with ionic and covalent crosslinks. J Appl Phys. 2010; 107 063509/1-063509/5. doi: 10.1063/1.3343265.

[34] Suzuki Y, Nishimura Y, Tanihara M, Suzuki K, Nakamura T, Shimizu Y, Yamawaki Y, Kakimaru Y. Evaluation of a novel alginate gel dressing: Cytotoxicity to fibroblasts in vitro and foreign-body reaction in pig skin in vivo J Biomed Mater Res. 1998; 39:317-22. doi.org/10.1002/(SICI)1097-4636(199802)39:2<317::AID-JBM20>3.0.CO;2-8

[35] Eiselt P, Lee KY, Mooney DJ. Rigidity of two-component hydrogels prepared from alginate and poly(ethylene glycol)-diamines. Macromolecules 1999;32:5561-6. doi.org/10.1021/ma990514m.

[36] Lee KY, Bouhadir KH, Mooney DJ. Controlled degradation of hydrogels using multifunctional cross-linking molecules. Biomaterials , 2004;25: 2461-6. doi.org/10.1016/j.biomaterials.2003.09.030.

[37] Lu MZ, Lan HL, Wang FF, Chang SJ, Wang YJ. Cell encapsulation with alginate and phenoxycinnamylidene-acetylated poly(allylamine). Biotechnol Bioeng 2000;70:479-83.doi.org/10.1002/1097-0290(20001205)70:5<479::AID BIT1>3.0.CO;2-E.

[38] Das M K, Senapati PC. Furosemide-loaded Alginate microspheres prepared by ionic cross-linking technique: Morphology and release characteristics. Indian J. Pharm. Sci. 2008, 70, 77-84. doi: 10.4103/0250-474X.40336.

[39] da Silva TL, Vidart JM, da Silva MG, Gimenes ML, Vieira MG. Alginate and Sericin: Environmental and Pharmaceutical Applications. In Biological Activities and Application of Marine Polysaccharides; InTech: Rijeka, 2017; pp 57-86. DOI: 10.5772/65257.

[40] Liu, X. D.; Yu, W. Y.; Zhang, Y.; Xue, W. M.; Yu, W. T.; Xiong, Y.; Ma, X. J.; Chen, Y.; Yuan, Q. Characterization of structure and diffusion behaviour of Ca-Alginate beads prepared with external or internal calcium sources. J. Microencapsul. 2002, 19, 775-782. doi.org/10.1080/02652040210000022743.

[41] Mahmoodi, NM. Magnetic ferrite nanoparticle–Alginate composite: Synthesis, characterization an binary system dye removal. J. Taiwan Inst. Chem. Eng 2013, 44,322-330. doi.org/10.1016/j.jtice.2012.11.014.

[42] Martinsen, A.; Skjåk-Braek, G.; Smidsrød, O. Alginate as immobilization material: I. correlation between chemical and physical properties of Alginate gel beads. Biotechnol. Bioeng. 1989, 33, 79-89. DOI: 10.1002/bit.260330111.

[43] Vandenbossche,G.M. R.; Remon,J.P. Influence of the sterilization process on alginate dispersions. J. Pharm. Pharmacol. 1993, 45 (16),484-486. https://doi.org/10.1111/j.2042-7158.1993.tb05582.x.

[44] Gombotz,W.R .; Wee,S.F. Protein release from alginate matrices. Adv. Drug Deliv. Rev. 1998, 31 (1-2), 267-285. DOI: 10.1016/s0169-409x(97)00124-5.

[45] Becker TA, Kipke DR, Brandon T. Calcium alginate gel: A biocompatible and mechanically stable polymer for endovascular embolization. J. Biomed. J Biomed Mater Res. 2001;54(1):76-86. DOI: 10.1002/1097-4636(200101)54:1<76::aid-jbm9>3.0.co;2-v.

[46] Stabler C, Wilks K, Sambanis A, Constantinidis. The effects of alginate composition on encapsulated bTC3 cells. Biomaterials 2001, 22 (11), 1301-1310. doi.org/10.1016/S0142-9612(00)00282-9.

[47] Shamkhani A, Duncan R. Radioiodination of alginate via covalently-bound tyrosinamide allows monitoring of its fate in vivo. J Bioact Compat Polym 1995;10:4-13.doi.org/10.1177/088391159501000102.

[48] Torres ML, Fernandez JM, Dellatorre FG, Cortizo AM, Oberti TG. Purification of alginate improves its biocompatibility and eliminates cytotoxicity in matrix for bone tissue engineering. Algal Research, 2019;40:101499, doi.org/10.1016/j.algal.2019.101499.

[49] Peppas NA, Buri PA. Surface interfacial and molecular aspects of polymer bioadhesion on soft tissues. J. Control Release. 1985; 2: 257-275. Doi.org/10.1016/0168-3659(85)90050-1.

[50] Serp D, Cantana E, Heinzen C, Stockar UV, Marison IW. Characterization of an encapsulation device for the production of monodisperse alginate beads for cell immobilization. Biotech. Bioeng. 2000; 70 (1):41-53. doi.org/10.1002/1097-0290(20001005)70:1<41::AID-BIT6>3.0.CO;2-U

[51] Gaserod,O, Jolliffe I.G, Hampson FC, Dettmar PW, Skjak-Braek G. The enhancement of the bioadhesive properties of calcium alginate beads by coating with chitosan. Int. J. Pharm. 1998; 175 (2): 237-246. doi.org/10.1016/S0378-5173(98)00277-4.

[52] Bernkop-Schnurch A, Kast CE, Richter MF. Improvement in the mucoadhesive properties of alginate by the covalent attachment of cysteine. J. Control Release. 2001;71(3):277-285. doi.org/10.1016/S0168-3659(01)00227-9.

[53] Dusel R, McGinity J, Harris MR, Vadino W A, Cooper J. Sodium alginate. In Handbook of Pharmaceutical Excipients; Published by American Pharmaceutical Society: USA and The Pharmaceutical Society of Great Britain: London. 1986; 257-258.

[54] Anu Shilpa, S. S. Agrawal; Alok R. Ray, Polymer Reviews. 2003;43(2):187 -221. DOI: 10.1081/MC-120020160.

[55] Mallikarjuna B, Madhusudana Rao K, Siraj S, Chandra Babu A, Chowdoji Rao K, Subha, MCS. Sodium alginate/poly (ethylene oxide) blend hydrogel membranes for controlled release of Valganciclovir hydrochloride. Designed Monomers and Polymers. 2013;16(2);151. doi.org/10.1080/15685551.2012.705503.

[56] Okunlola A, Patel R.P, Odeku OA. Evaluation of freeze-dried pregelatinized Chinese yam (Dioscorea oppositifolia) starch as a polymer in floating gastroretentive metformin microbeads. Journal of Drug Delivery Science and Technology. 2010; 20 (6): 457-465. doi.org/10.1016/S1773-2247(10)50079-0

[57] Nayak PL, Debasish Sahoo. Chitosan-alginate composites blended with Cloisite 30B as a novel drug delivery system for anticancer drug paclitaxel. International Journal of Plastics Technology.2011; 15(1): 68. DOI:10.1007/s12588-011-9000-6.

[58] Zeng WM. Oral controlled release formulation for highly water-soluble drugs: Drug–sodium Alginate–xanthan gum–zinc acetate matrix. Drug Development and Industrial

Pharmacy.2004; 30 (5):491-495. doi.org/10.1081/DDC-120037479

[59] Cui-Yun Yu, Bo-Cheng Yin, Wei Zhang, Si-Xue Cheng, Xian-Zheng Zhang, Ren-Xi Zhuo. Composite microparticle drug delivery systems based on chitosan, alginate and pectin with improved pH-sensitive drug release property. Colloids and Surfaces B: Biointerfaces, 2009;68 (2):245.doi.org/10.1016/j.colsurfb.2008.10.013.

[60] Nayak AK, Pal. D. Development of pH-sensitive tamarind seed polysaccharide-alginate composite beads for controlled diclofenac sodium delivery using response surface methodology. International Journal of Biological Macromolecules. 2011; 49(4):784. doi: 10.1016/j.ijbiomac.2011.07.013.

[61] Ramesh Babu V, Malladi Sairam, Kallappa M, Hosamani, Tejraj Aminabhavi M. Preparation of sodium alginate–methylcellulose blend microspheres for controlled release of Nifedipine.Carbohydrate Polymers. 2007; 69(2): 241-250. doi.org/10.1016/j.carbpol.2006.09.027.

[62] Oya Şanl, Nuran Ay , Nuran Işıklan. Release characteristics of diclofenac sodium from poly (vinyl alcohol)/sodium alginate and poly(vinyl alcohol)-grafted-poly(acrylamide)/sodium alginate blend beads. European Journal of Pharmaceutics and Biopharmaceutics. 2007; 65(2): 204. doi: 10.1016/j.ejpb.2006.08.004.

[63] Rajesh N, Siddaramaiah, Gowda DV, Somashekar CN. Formulation and evaluation of biopolymer based transdermal drug delivery, International Journal of Pharmacy Pharmaceutical Sciences, 2010, 2, Suppl 2:142-148. https://innovareacademics.in/journal/ijpps/Vol2Suppl2/569.pdf.

[64] Rohit chaudhary, MD. Shamim Qureshi, Jitendra Patel, Uttam prasad panigrahi, Giri, I.C. Formulation, Development and in-vitro evaluation of mucoadhesive buccal patches of methotrexate. International Journal of Pharma Sciences and Research.2010; 1(9):357.

[65] Lee DW, Hwang SJ, Park JB, Park HJ. Preparation and release characteristics of polymer-coated and blended alginate microspheres. Journal of Microencapsulation. 2003; 20(2): 179-192. doi.org/10.3109/02652040309178060.

[66] Mukesh Shinde B, Ganesh Dama. Studies on development of in situ gelling system of ciprofloxacin. Journal of Advanced Drug Delivery.2014;1(1),2. http://www.jadd.in/Content/Paper/201919575310000112019195753100001.pdf.

[67] Lee KY, Peters MC, Mooney DJ. Comparison of vascular endothelial growth factor and basic fibroblast growth factor on angiogenesis in SCID mice. J Control Release 2003; 87:49-56. DOI: 10.1016/s0168-3659(02)00349-8.

[68] Silva EA, Mooney DJ. Effects of VEGF temporal and spatial presentation on angiogenesis. Biomaterials 2010; 31: 1235-1241. doi: 10.1016/j.biomaterials.2009.10.052

[69] Wells LA, Sheardown H. Extended release of high pI proteins from alginate microspheres via a novel encapsulation technique. Eur J Pharm Biopharm 2007;65: 329-335. DOI:10.1016/j.ejpb.2006.10.018.

[70] Gao CM, Liu MZ, Chen SL, Jin SP, Chen J. Preparation of oxidized sodium alginate-graftpoly((2-dimethylamino) ethyl methacrylate) gel beads and in vitro controlled release behavior

of BSA. Int J Pharm. 2009; 371:16-24. doi: 10.1016/j.ijpharm.2008.12.013.

[71] Yu CY, Yin BC, Zhang W, Cheng SX, Zhang XZ, Zhuo RX. Composite microparticle drug delivery systems based on chitosan, alginate and pectin with improved pH-sensitive drug release property," Colloids and Surfaces B: Biointerfaces, 2009; 68(2):245-249. DOI: 10.1016/j.colsurfb.2008.10.013

[72] Chen AZ, Chen MY, Wang SB, Huang XN, Liu YG, Chen ZX. Poly(L-histidine)-chitosan/alginate complex microcapsule as a novel drug delivery agent. Jol of App Polymer Science. 2012; 124(5): 3728-3736. Doi.org/10.1002/app.35371.

[73] Alejandro Sosnik, Alginate Particles as Platform for Drug Delivery by the Oral Route: State-of-the-Art International Scholarly Research Notices. 2014, Article ID 926157. http://dx.doi.org/10.1155/2014/926157.

[74] AlgiSite M™. http://www.smith-nephew.com/professional/products/advanced-woundmanagement/algisite-m.

[75] SMTL Dressings Datacard.: http://www.dressings.org/Dressings/comfeel-plus.html.

[76] KALTOSTAT Calcium Sodium Alginate Dressing. https://fsastore.com/KALTOSTATCalcium-Sodium-Alginate-Dressing-3-x-4-34-Box-of-10 P23356.aspx.

[77] Sorbsan Flat. http://www.aspenmedicaleurope.com/specialist_wound_car/sorbsan-flat/

[78] HYALOGRAN, Biodegradable Wound Dressing. Available online: http://www.anikatherapeutics.com/products/dermal/hyalogran/

[79] Algivon. Available online: http://www.advancis.co.uk/products/activon-manuka-honey/algivon.

[80] FIBRACOL™ Plus. Collagen Wound Dressing with Alginate. http://www.woundsource.com/product/fibracol-plus-collagen-wound-dressing-alginate.

[81] Szekalska M, Puciłowska A, Szyma´ nska E, Ciosek P, Winnicka K. Alginate: Current Use and Future Perspectives in Pharmaceutical and Biomedical Applications. Int. J. Polym. Sci. 2016; 8:1-17. Doi.org/10.1155/2016/7697031.

[82] https://www.directmedicalinc.com/catalog/product/1514/derma-sciences-medihoney-calcium-alginate-dressing-31045/

[83] https://www.lodaites.com/blogs/news/algues-cosmetique?lang=en

[84] https://www.makingcosmetics.com/ICE-Alginate_p_129.html?locale=en

[85] https://kimica-algin.com/alginate/application/

[86] https://glidewelldental.com/solutions/removable-prosthesis/partials/flexible-partials

[87] https://www.pinterest.com/LVSmileCenter/dentures/

[88] https://www.iroalginate.com/app-welding.htm.

[89] Maria de Lourdes Bastos, Georges Bories, Andrew Chesson, Pier Sandro Cocconcelli, Gerhard Flachowsky, Boris Kolar, Maryline Kouba, Marta L_opez-Alonso, Secundino L_opez Puente, Alberto Mantovani, Baltasar Mayo, Fernando Ramos, Maria Saarela, Roberto Edoardo Villa, Robert John Wallace, Pieter Wester, Anne-Katrine

Lundebye, Carlo Nebbia, Derek Renshaw, Safety and efficacy of sodium and potassium alginate for pets, other nonfood producing animals and fish. EFSA Journal 2017; 15(7):4945.

[90] Kent Lanter, Brenda de Rodas,Bill L. Miller, Gary E. Fitzner, US Patent US 8092853 B2., Jan 2012. doi: 10.2903/j.efsa.2017.4945.

Chapter 3

Heavy Metal Removal by Alginate Based Agriculture and Industrial Waste Nanocomposites

Shivangi Omer

Abstract

The use of biopolymers and nonliving organisms as sorbents is one of the most promising techniques because they contain several functional groups which show different affinities towards various metal ions. Alginate is naturally occurring anionic biopolymer extracted from brown algae. It also contains numerous applications in biomedical science and engineering due to its favorable properties and ease of gelatin. This chapter represents a overview based on alginate based agriculture and industrial waste nanocomposites and found that limited studies are reported for combination of alginate with industrial/agriculture waste in nanoscalic material so far, but this review study enlightening the several studies based on nanocomposite combinations of alginates and biopolymers and these biopolymers can also be derived from various agro/industrial waste by simple chemical and mechanical methods. So, we should work on the formulation of alginates agro/industrial waste nanocomposites. Preparation of alginate nanomaterials with agriculture/industrial waste constituents confirms its effectiveness in water purification. In the environment, we can control its reutilization by desorption studies. Another advantage is that it can be transform from nanoparticles to nano polymeric films and support to batch adsorption process to fixed bed column in form of large-scale application.

Keywords: Alginate, agriculture waste, industrial waste, nanocomposite, adsorption, heavy metals

1. Introduction

Alginate is naturally occurring anionic polymer typically obtained from brown seaweed and commercially available alginate extracted from brown algae (Phaeophyceae), including *Laminaria hyperborea*, *Laminaria digitata*, Laminaria japonica, *Ascophyllum nodosum*, and *Macrocystis pyrifera*. Alginate is calcium, magnesium, and sodium salt of alginic acid which is a linear structure of heteropolysaccharide and composed of d-mannuronic acid and l-guluronic acid.

It is biodegradable, biocompatible, renewable, nontoxic and moderately efficient because of their low molecular weight, high shear stability and easily available from reproducible farm and forest resources. It also contains numerous applications in biomedical science and engineering due to its favorable properties and ease of gelatin [1]. As its utilization in agriculture application, it functions as soil conditioner. Alginate is natural materials that can absorb large amount of water, as much as hundreds of times of their own mass and can form ionic and nonionic moisture-holding hydrogels which increase soil water retention capability. So, alginate is excellent superabsorbent or water retaining material and can serve as a good carrier to uptake metal ions from aqueous solution [2].

The use of biopolymers and nonliving organisms as sorbents is one of the most promising techniques because they contain several functional groups which show different affinities towards various metal ions. As we say earlier, alginate is component of outer cell of brown algae but commercially it exist as sodium alginate (SA) which shows viscous nature after dissolving in water, but when it is utilized for removal of metal ions, it is usually prepared as Calcium Alginate. Biopolymers contains many potential binding sites such as carboxylate, amine, phosphate, sulfate, hydroxyl and other chemical functional groups. And due to availability of these free binding sites enhance surface charges. As a result, increase uptake of metals ions. Binding with these free active sites largely depend on pH of the aqueous solution and various metal ions shows different adsorption capacity at different pH scale. As a point of view of alginates chemical structure, Sodium alginate (SA) is rich in hydroxyl (OH) and carboxyl (COOH) functional groups and extra negatively charged sites, which are the foremost groups involved in heavy metal adsorption. Alfaro-Cuevas-Villanueva et al. evaluated the removal of lead and cadmium using calcium alginate beads as biosorbents under the influence of various pH range and temperature [3].

But a potential problem with alginates derivatives are their poor strength and flexibility towards high hydrophilicity. Which can be overcome by chemical modification with various biobased polymers such as cellulose, starch, chitosan, xylene, lignin. It will not only modify properties related with basic structure but also will enhance biocompatibility and biodegradability of compound. Of course, hydrophilicity of these hybridized copolymers introduces water fluxes with greater adsorption capacity higher than most of the synthetic polymers which also a green approach to environment.

In terms of operational costs and ease to use, adsorption is the most promising techniques among all the techniques reported in various literature. Adsorption allowing the treatment waters gives rise to a very rich bibliography through a very great variability of adsorbents.

Since this chapter include contents about alginate-based agriculture/industrial waste nanocomposites, there is a need to brief about agriculture/industrial waste.

2. Agriculture waste/industrial waste

2.1 Agriculture waste

Various wastes related agriculture cropping, processing or harvesting are called as 'Agriculture waste'. Agriculture waste are bioactive compounds which are rich in nutrients such as cellulose, lignin, xylene and fibers. Agriculture based industries produces huge number of residues as waste product every year which are dumped to the open ground area and release to environment without proper disposal. But these residues can be used utilized as alternate source in various application such as biogas, biofuels, as an adsorbent for heavy metal, in mushroom cultivation and as raw material in various industries that can help to reduce the production cost and relived the pollution load of environment. Various organizations defined agro-industrial waste by different ways.

Obi et al. reported, Agricultural wastes are defined as the residues from the growing and processing of raw agricultural products such as fruits, vegetables, meat, poultry, dairy products, and crops [4]. They are the unwanted outputs of production and processing of agricultural products that may contain material that can benefit man but whose economic values are less than the cost of collection, transportation, and processing for beneficial use [5–7].

Their composition of agriculture waste will depend on the system and type of agricultural activities takes place after the process of cropping and they can be in various forms such as liquids, slurries, or solids. Agricultural waste are also called as agro-waste is which comprised of natural, animal waste (manure, animal carcasses), food processing waste (only 20% of maize is canned and 80% is waste), crop waste (corn stalks, sugarcane bagasse, drops and culls from fruits and vegetables, prunings) and hazardous and toxic agricultural waste (pesticides, insecticides and herbicides) etc. [4, 5]. **Figure 1** illustrates a simple classification of agriculture waste based on its source.

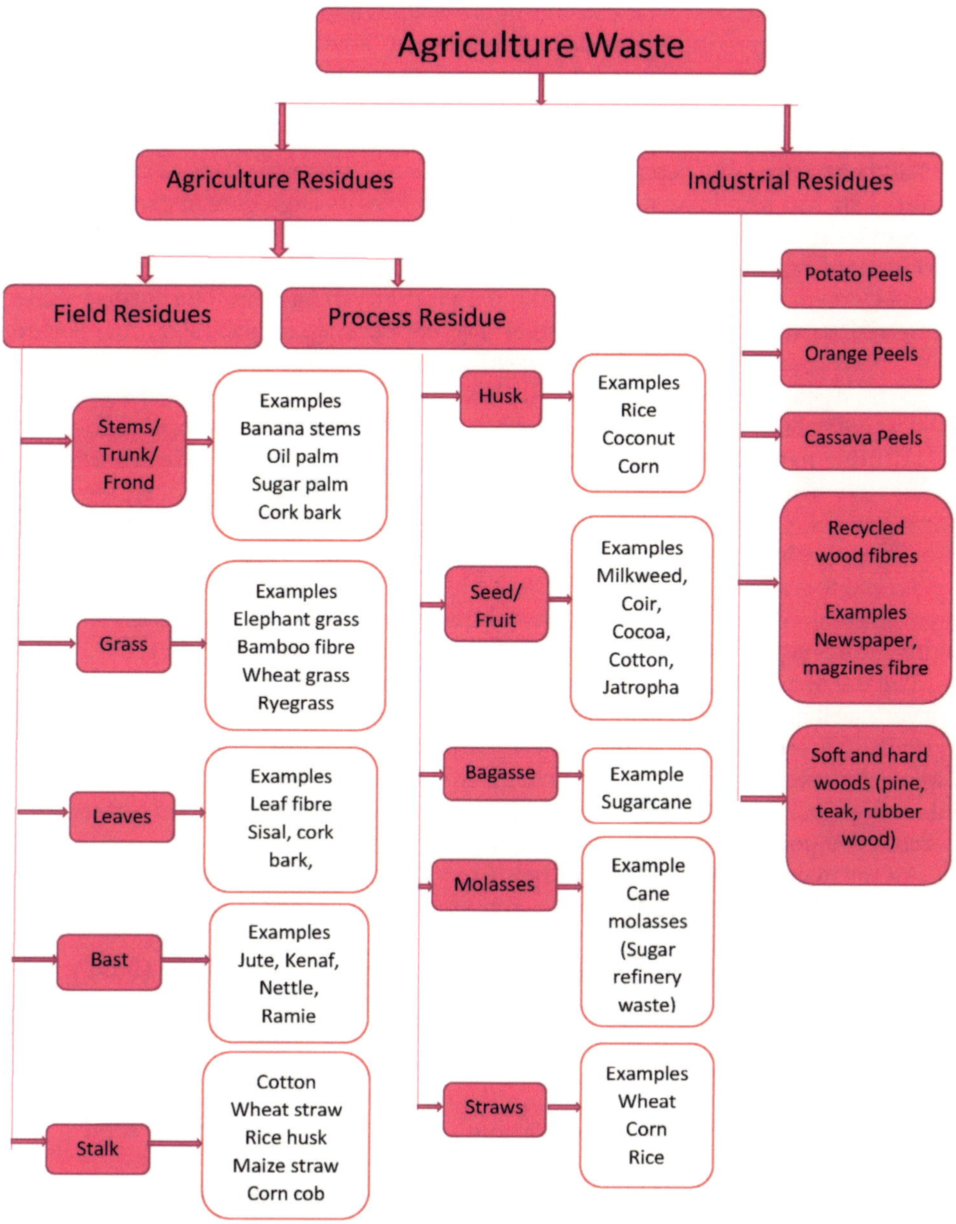

Figure 1.
Classification of agriculture waste based on source.

2.2 Industrial waste

We have millions of factories, mills, industries all around the world that use several types of raw materials to manufacture different type of products, goods, cosmetics, pharmaceuticals or other things. So, of course these all-manufacturing sites generate a vast amount of material which is rendered useless and called as 'Industrial waste'.

So, Industrial waste is defined as waste generated by manufacturing or industrial processes. The types of industrial waste generated include cafeteria garbage, dirt and gravel, masonry and concrete, scrap metals, trash, oil, solvents, biomass ash, discarded tires, chemicals, weed grass and trees, wood and scrap lumber, and similar wastes [8–12].

These all industrial wastes containing hazardous and nonhazardous components which either released into natural water bodies, or burnt or throw in open ground area without concerning about environmental elements. So, we can utilize all these wastes generated from small- or large-scale industries for so many applications or we can utilize all these scrapes by modification in chemical structures.

Saikia and Brito [9], reported the application of industrial waste as building material such as plastic waste in concrete, coal ash as an aggregate in concrete and rubber tire as a filler in concrete. He also enlightened on utilization of alloy slags in various applications. Dash et al. [13], reported a sustainable use of industrial waste as replacement of fine aggregate for the preparation of concrete. They utilize some industrial waste like foundry sand, steel slag, copper slag, furnace slag, coal bottom ash, ferrochrome slag, palm oil clinker etc. as a replacement of sand or aggregate in concrete. Aggarwal and Siddique [14], also utilize bottom ash and waste foundry sand as partial replacement of fine aggregates which gives us a fine option for substituting natural fine aggregate with industrial by-products. There are several studies in which industrial byproducts are used as adsorbent for the sorption of harmful dyes, pigments and heavy metals present in waste water streams. Anionic dyes such as ethyl orange, metanil yellow and acid blue removal were investigated by inorganic wastes such as blast furnace, sludge, slag from steel plants and by organic material prepared from carbon slurry of fertilizer. Various industrial waste or by-products produce by various industries are summarize in **Table 1**.

The use of agriculture waste as adsorbent towards many heavy metal ions are very well reported in various studies. The major chemical composition of agricultural residues with their properties and functions are depicted in **Figure 2**. Lignocellulosic biomass based on agriculture residues mainly composed of cellulose, lignin, hemicellulose, starch, ash etc. All these residues contain different percent composition of these components.

All the major components with their function and properties are shown in **Figure 2** and percent weight of these components consisted by different agriculture waste are shown in **Table 2** [15–18].

Huq et al. [19], efficiently transformed nanocrystalline cellulose (NCC) into reinforced alginate-based nanocomposite film. This study also gives a strong support for the utilization of agriculture waste as a rich source of biopolymer like cellulose. In this study NCC is prepared form wood pulp by acid hydrolysis and activated by treatment with sulfuric acid and then incorporated into nanoscalic alginate structure to produce a renewable and biodegradable film for food packaging applications.

All the agriculture wastes contain cellulose and hemicellulose as main part of its chemical structure, these all-agriculture residues are also referred as "Lignocellulosic mass". The cellulose and hemicellulose are polysaccharides which cover two-third part of the lignocellulosic biomass, which can be bioconverted in

S. N.	Industrial sector	Description	Typical waste
1.	Mining and quarrying	Extraction, beneficiation, and processing of minerals	Solid rock, slag, phosphogypsum, muds, tailings
2.	Energy	Electricity, gas, steam, and air-conditioning supply	Fly ash, bottom ash, boiler slag, particulates, used oils, sludge
3.	Manufacturing	Chemical	Spent catalyst, chemical solvents, reactive waste, acid, alkali, used oils, particulate waste, ash, sludge
4.	Food	Food	Plastic, packaging, carton
5.	Textile		Textile waste, pigments, peroxide, organic stabilizer, alkali, chemical solvents, sludge, heavy metals
6.	Construction	Construction, demolition activity	Concrete, cinder blocks, gypsum, masonry, asphalt, wood shingles, slate, metals, glass, and plaster
7.	Waste/water services	Water collection, treatment, and supply	Spent adsorbent, sludge

Table 1.
Industrial waste produced from various industries.

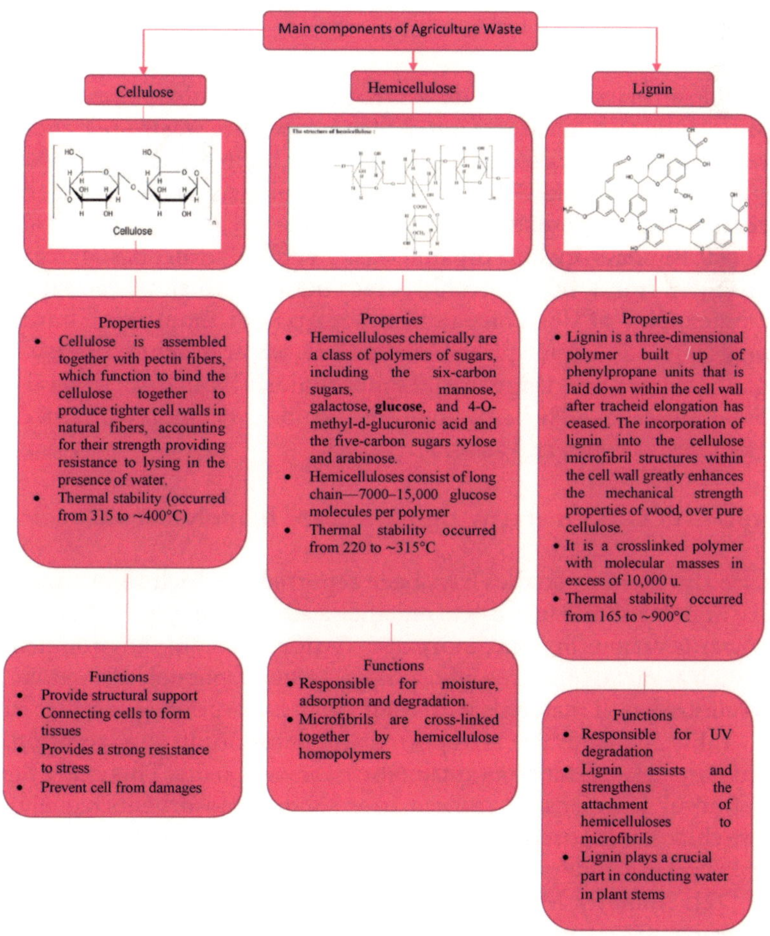

Figure 2.
Main components of agriculture waste.

S. N.	Agriculture residues	Cellulose (wt%)	Hemicellulose (wt%)	Lignin (wt%)	Ash (wt%)
1.	Wheat Straw	43.2	34.1	22.0	1.8
2.	Sugar palm fiber	43.88	7.24	33.24	1.01
3.	Rice Husk	41.05	17.63	18.82	—
4.	Sugarcane bagasse	43.6	27.7	27.7	—
5.	Banana fiber	7.5	74.9	7.9	0.01
6.	Coconut husk	43.44	0.25	45.84	2.22

Table 2.
Percent composition of agriculture residues.

fermentable sugars like hexoses and pentoses under the action of cellulase enzymes. For this enzymatic hydrolysis is required and this rate of hydrolysis and cellulase enzyme activity is improved by presence of alginate. It enhances enzyme immobilization capacity which can further improve the stability and hydrolytic efficiency of cellulase. And the presence of Fe_3O_4/Alginate nanocomposite enhanced the production of reducing sugars from lignocellulosic mass – rice straw has also been shown by Srivastava et al. [20].

2.3 Lignocellulosic biomass

Lignocellulosic mass are the materials that contain cellulose as main content of its composition and rich in various nutrients. Agriculture waste are rich in cellulosic material so, these are also called as 'Lignocellulosic Biomass', as it is produced by various agro based industries. These materials are biodegradable neither toxic nor hazardous and have potential to play a vital role in various applications. Lignocellulosic Biomass' include organic wastes such as corncob, sugarcane bagasse, sugar palm (fiber, frond, bunch, trunk), areca nut husk fiber, wheat straw fiber, soy hull fiber, pineapple leaf fiber, oil palm (mesocarp fiber, empty fruit bunch, frond), rubber wood thinning, curaua fiber, banana fiber, water hyacinth fiber, wheat straw, sugar beet fiber, etc. (**Figure 3**). Lignocellulosic biomass in general consists of 35–55% cellulose, 25–40% hemicellulose, and 15–25% lignin with small percentage of extractives, protein, and ash [21].

2.4 Adsorption of heavy metal by alginate-based biopolymer nanocomposites

There are also so many studies have been reported in which alginate structure plays a very important role in order to enhance a adsorption capacity of synthesized material towards various organic, inorganic, Cationic, anionic environmental pollutants such as metal ion, dyes, phosphates, toluene, phenol, nitrates etc. various alginate structure based materials with their batch adsorption studies are summarized in **Table 3**. Lakouraj et al. reported an intense study on adsorption capacity of SA based nanogel and superparamagnetic nanocomposite. This study represents synthesis of novel, inexpensive, and eco-friendly nanobiosorbents namely, magnetic tetrasodium thiacalix [4] arene tetrasolfonate supported sodium alginate (TSTC [4] AS-s-SA) and utilized for sorption of Cu(II), Cd(II), Pb(II), Co(II), Ni(II) and Cr(II) [22]. Similarly, Mittal et al. [23], copolymerized alginate with methyl methacrylate magnetic nanocomposite for the sorption of Pb(II) and Cu(II) metal ions. Phiri et al. [24], fabricated novel magnetic nanocomposite alginate beads in which alginate beads impregnated with a new combination - bentonite, zeolite,

Figure 3.
Examples of lignocellulosic mass.

S. N.	Agro/industrial waste	pH	Other conditions	Adsorbates	Source
1.	Calcium alginate beads	6.0 to 7.0	25°C	Pb^{2+} and Cd^{2+}	[3]
2.	Tetrasodium thiacalix[4] arene tetrasulfonate (TSTC[4]AS-s-SA) nanogel	7	—	Cu(II), Cd(II), Pb(II), Co(II), Ni(II) and Cr(III)	[22]
3.	Poly (methyl methacrylate)-grafted alginate/Fe_3O_4 nanocomposite	5	50°C	Pb(II) and Cu(II)	[23]
4.	Alginate with bentonite, zeolite, activated charcoal and aluminum – zinc layered double hydroxides beads		—	Phosphate, copper, toluene	[24]
5.	Polyvinyl alcohol/sodium alginate (PVA/SA) beads	6.0 to 7.0		Pb^{2+}, Cd^{2+}, Sr^{2+}, Cu^{2+}, Zn^{2+}, Ni^{2+}, Mn^{2+} and Li^{2+}	[25]
		5		Fe^{3+} and Al^{3+}	
6.	Cobalt ferrite - alginate nanocomposite	3–6	60 min	binary dye effluent	[26]
7.	Polyvinyl alcohol/graphene oxide-sodium alginate nanocomposite hydrogel		—	Pb^{2+}	[27]

Table 3.
Batch adsorption studies of various alginate-based biopolymer nanocomposite as adsorbent.

activated charcoal and aluminum – zinc layered double hydroxides was synthesized for removal of phosphate, copper and toluene with complete adsorption of these organic pollutants.

Karkeh-Abadi et al. [28], reported the synthesis of sodium alginate hydroxyapatite-CNT nanocomposite beads used as adsorbents of Cobalt ions from an aqueous solution. While Gokila et al. [29], show the performance of Chitosan and Alginate nanocomposites and proved as an excellent material for sequestration of Cr(VI) ions from waste water.

Enhancement in adsorption capacity by calcium alginate gel beads towards Pd(II) ion is also reported by Cataldo et al. [30]. In this study, two type of clay material Montmorillonite and Laponite utilized for synthesizing hybrid materials on the base of Ca-alginate.

The deprotonation of carboxylate ions present in alginate structure plays a very important role its enhanced binding ability with metal ions which in turn leads to greater adsorption capacity.

Esmat et al. [31], discussed alginate-based nanocomposites for the efficient removal of heavy metal ions. They prepared alginate-based nanocomposites by synthesizing cobalt ferrite nanoparticles (CF) and titanate nanotubes (T) and utilized as potential adsorbents for efficient removal of Cu^{2+}, Fe^{3+} and As^{3+} ions from water. This study shows that the presence of divalent cations such as Ca^{2+}, alginate transforms a gel form polymer matrix very easily which is easier to handle than powder materials and easily separable for desorption studies. The presence of alginates in nanostructures supports complex formation with metal ions present in aqueous solution which also increase its adsorption capacity to better extent.

Magnetic core shell nanoadsorbent using alginate base also have been coming forth as a possible option to remove organic and inorganic pollutants. Cobalt-ferrite nanoparticles embedded in alginate polymer matrix contain highly reactive surface and also allow fast magnetic separation after the adsorption process [26].

2.4.1 Silica based alginate bio-nanocomposite

There are various studies potrating silica as potential energy source for biofuels production, biochars, catalytic reforming, pollutant removal, soil remediation, waste water treatment and gas purification by means of adsorption, catalysis and another integrated process [32–34]. Nanoengineered Silica particles are widely used for biomedical purposes, in cosmetic products, food industry, automobiles, paints etc. Even rice husk silica waste is also used for the preparation of porous materials [35]. Waste material is a potential alternative source of silica. Extraction of silica from various agriculture as well as industrial waste are also very well reported [36–38]. Raja et al. [39], efficiently discussed the extraction methods of silica from agrowaste. Shen [34], enlightened various sustainable application for Rice husk silica derived nanomaterials. Sapawe et al. [40], synthesized silica from six different material as agriculture waste - sugarcane bagasse, bamboo culm, bamboo leaf, corncob, banana leaf and cigarette butt by sol gel method i.e., most common method to prepare polymeric network of gel. Salamaa et al. [41], crosslinked alginate with silica and zinc oxide in which silica is extracted from rice husk by conventional method and hybrid nano material is prepared by treating with alginate as matrix supporting material and investigated for antibacterial properties. However, adsorption property of silica alginate nanocomposites is not investigated yet but it has been shown former studies that alginate has played a very important role in enhancing the adsorption phenomena and as like alginate, silica also showed a potential adsorption capacity towards heavy metal ions. So, by combining these two compounds would result in a very efficient adsorbent nanocomposite material.

2.4.2 Carbon based alginate bio-nanocomposite

Carbon based nano adsorbents such as carbon nanotubes, graphene, nanodiamond, fullerenes, nanosized carbon allotropes have attracted tremendous interest for sorption of metal ion removal from wastewater due to their extraordinary performance with complete adsorption capacity and easy operational use but not low cost effective. By their chemical properties these derivatives contain high aspect ratios, surface areas, porosities and reactivities. Moreover, they exhibit high electrical and thermal conductivity, great mechanical property and structural stability in extreme conditions such as high temperature, strong acidic/alkaline conditions. Due to their high hydrophobicity and small sized, it is difficult to recollect from aqueous solution. To overcome such disadvantages, many studies had been reported to modify chemical structures by grafting with biopolymeric network [42]. Yu et al. [27], prepared polyvinyl alcohol/graphene oxide-sodium alginate nanocomposite hydrogel through in situ crosslinking for Pb^{2+} removal which reveals that intercalation of GO structure in alginate based polymeric network increase metal ion complexation because of its abundant functional groups, such as hydroxyl, carboxyl, carbonyl, and epoxy groups on the surfaces. Similarly, Karkeh-Abadi et al. [28] incorporated functionalized CNT in the network of sodium alginate-based nanocomposite beads on the removal of Co(II) ions from aqueous solutions. Here, CNTs are promising synthetic polymers because of their large surface area, greater chemical reactivity, high aspect ratio, less chemical mass and impact on the environment. Yadav et al. [43] fabricated Novel Magnetic/Activated Charcoal/β - Cyclodextrin/Alginate Polymer Nanocomposite and utilized for elimination of methylene blue dye. As activated charcoal has been used as a common "universal" adsorbent, since it is a simple, safe, high surface area and high adsorption capacity material. Konwar and Chowdhury [44] studies property relationship of alginate and alginate carbon dots nanocomposite with different bivalent and trivalent metal ions such as - Ca^{2+}, Ba^{2+}, Cu^{2+}, Fe^{3+} and Al^{3+}. Carbon dots were prepared from 'Assam Tea' and improved the properties like UV blocking, thermostability and mechanical strength of these prepared biopolymeric films. Ma et al. [45] reported enhanced adsorption for removal of antibiotics by carbon nanotubes/graphene oxide/sodium alginate triple-network nanocomposite hydrogels in aqueous solutions.

All these above studies incorporated carbon derivatives with alginate structure and resultant material were utilized for adsorption of metal ions in different forms. Agriculture as well as industrial waste are also key source of carbon derivatives. Various studies have been reported in which these carbon derivatives are extracted from agriculture/industrial residues. Mohan et al. [46], develop activated carbon from coconut shell for removal of pyridine derivatives. Kerdnawee et al. [47] discussed the advancement of carbonaceous nanomaterial production from industrial waste in detail. This study enlightened several studies reported for synthesis of CNT using industrial waste such as – chemical process waste, petroleum refining process waste, plastic waste, automobile waste, because these wastes contains high amount of CH4, C_2H_6, C_2H_4, C_2H_2, CO, CH_3OH and so on, would involve as carbon source for productions of carbonaceous material. Somanathan et al. [48] reported graphene oxide synthesis via single step reforming of sugar cane bagasse. It is green approach for synthesis of GO. There is also another review study which is reported for production of activated carbon from agriculture waste. This detailed study is collection of various studies based for production of activated carbon from agriculture waste such as – pineapple peels, rice husks, cotton stalk, orange peel, tobacco stems, bamboo, waste tea, industrial waste lignin, pine wood powder etc., by microwave method [49]. Ensuncho-Muñoz and Carriazo [50], reported for e preparation and characterization of carbonaceous materials obtained from three types of vegetable wastes provided by agricultural industries i.e.; coconut husk, corn cob and rice husk and utilized as adsorbent of azo dyes.

So, from all the above studies we can conclude that carbonaceous material can be produce from agriculture as well as industrial waste. I also reported for the studies based on preparation of alginate nanocomposites with different form of carbonaceous materials. So these studies evident that carbon based alginate nanocomposites can be synthesized by utilizing agriculture and industrial waste. There are also various methods like sol gel method, microwave heating method, modified hummers method, calcination method, grafting and copolymerization reaction with the help of crosslinkers are also come into consideration for preparation. Of course, no study has been found to be reported which involve direct linkage between alginate with carbonaceous derivatives obtained from agriculture/industrial waste and utilized for removal of heavy metals from wastewater, so there is research gap but definitely it could result into a very efficient adsorbent material.

3. Conclusion

So, utilization of alginate as biopolymer in nanocomposites is beneficial from every side of its applications. It is biodegradable, biocompatible, renewable, nontoxic and moderately efficient. That is why it contains numerous applications in biomedical science and engineering, agriculture, drug delivery, enzyme mobilization, heavy metal removal from industrial effluents, as functional food ingredient, mineralization of organic pollution, emulsifiers, consistency enhancers, and thickening agents in cosmetic formulas etc. Limited studies are reported for combination of alginate with industrial/agriculture waste in nanoscalic material so far, but this review study enlightening the several studies based on nanocomposite combination of alginates and biopolymers and these biopolymers can also be derived from various agro/industrial waste by simple chemical and mechanical methods. So, we should work on the formulation of alginates agro/industrial waste nanocomposites. Preparation of alginate nanomaterials with agriculture/industrial waste constituents confirms its effectiveness in water purification. In the environment, we can control its reutilization by desorption studies. Another advantage is that it can be transform from nanoparticles to nano polymeric films and support to batch adsorption process to fixed bed column in form of large-scale application. Apart from that utilization of biological materials as adsorbents for elimination of pollutants minimize the process price substantially and make adsorption method more ecological and feasible.

Author details

Shivangi Omer
School of Applied and Life Sciences, Uttaranchal University, Dehradun, India

*Address all correspondence to: shivangiomer@gmail.com

References

[1] Lee, K. Y, Mooney, D. J : Alginate: properties and biomedical applications. Progress in polymer science. 2012; *37*(1):106-126.

[2] Pereira, L., & Cotas, J. Introductory Chapter: Alginates-A General Overview. Alginates-Recent Uses of This Natural Polymer: intechopen; 2020. DOI: 10.5772/intechopen.88381

[3] Alfaro-Cuevas-Villanueva, R., Hidalgo-Vázquez, A. R., Cortés Penagos, C. D. J., Cortés-Martínez, R. Thermodynamic, kinetic, and equilibrium parameters for the removal of lead and cadmium from aqueous solutions with calcium alginate beads. The scientific world journal, 2014.

[4] Obi, F. O., Ugwuishiwu, B. O., & Nwakaire, J. N. Agricultural waste concept, generation, utilization and management. Nigerian Journal of Technology. 2016*; 35*(4):957-964.

[5] Sadh, P. K, Duhan, S, & Duhan, J. S. Agro-industrial wastes and their utilization using solid state fermentation: a review. Bioresources and Bioprocessing. 2018; *5*(1): 1.

[6] Omer, S., Singh. A. and Upadhyay S. Synthesis of Chitosan-g-Biomass Ash/Graphene Oxide Nanocomposite for the Removal of Copper and Chromium from Industrial Waste Water. Clay Research. 2019[a]; *38*(1):19-28.

[7] Omer, S., Singh, A., Sasmal, R., & Singhal, A. Adsorption studies on chitosan and guar gum nanocomposite by application of agriculture waste–A review. International Journal of Research and Analytical Reviews 2019[b];*6*: 1059-1062.

[8] Mittal, A., Ahmad, R., & Hasan, I. Poly (methyl methacrylate)-grafted alginate/Fe3O4 nanocomposite: synthesis and its application for the removal of heavy metal ions. Desalination and Water Treatment. 2016; 57(42): 19820-19833.

[9] Saikia, N., & de Brito, J. (2009). Use of industrial waste and municipality solid waste as aggregate, filler or fiber in cement mortar and concrete. Advanced Material Science and Research. 2009; *3*: 65-116.

[10] Singh, A. and Omer. S. Adsorption of Heavy metal using Biomass ash and Guar gum Nanocomposite from waste water effluent – A review', SSRN- Elsevier. 2018; 1: 3.

[11] Omer, S., Anand, S., Singh, S. K., Singh, A. Review Study on Waste Tire as an Adsorbent to Remove Toxic Metals from Waste Water. International Journal of Pharmaceutical and Medicinal Research. *2020; 8*(1): 1-5.

[12] Omer, S. & Singh, A. Synthesis of Guar Gum -g-Coconut husk (SnO_2-SiO_2) Nanocomposite For Removal of Cu(II) and Cr(VI) From Waste Water. International Journal of Environment and Waste Management. 2022 (in press); 1(1): 1. DOI:10.1504/ijewm.2022.10036122.

[13] Dash, M. K., Patro, S. K., & Rath, A. K. Sustainable use of industrial-waste as partial replacement of fine aggregate for preparation of concrete–A review. International Journal of Sustainable Built Environment. 2016; 5(2): 484-516

[14] Aggarwal, Y., & Siddique, R. Microstructure and properties of concrete using bottom ash and waste foundry sand as partial replacement of fine aggregates. Construction and Building Materials. 2014; 54: 210-223.

[15] Faisal, Z. T., Amri, F., & Tahir, I. (2010). Effect of maleic anhydride polypropylene on the properties of coconut shell filled thermoplastic elastomeric olefin composites. Indonesian Journal of Chemistry. 2010; 10(2): 156-161.

[16] Verma, D., & Gope, P. C. The use of coir/coconut fibers as reinforcements in composites. In Biofiber Reinforcements in Composite Materials. 2015: 285-319.

[17] Gregory, A. and Bolwell, G.P. Hemicelluloses. Comprehensive Natural Products Chemistry. 1999; 3: 599-615.

[18] Tian, S. Q., Zhao, R. Y., & Chen, Z. C. Review of the pretreatment and bioconversion of lignocellulosic biomass from wheat straw materials. Renewable and Sustainable Energy Reviews. 2018; 91: 483-489.

[19] Huq, T., Salmieri, S., Khan, A., Khan, R. A., Le Tien, C., Riedl, B., ... & Lacroix, M. Nanocrystalline cellulose (NCC) reinforced alginate based biodegradable nanocomposite film. Carbohydrate polymers. 2012; 90(4): 1757-1763.

[20] Srivastava, N., Singh, J., Ramteke, P. W., Mishra, P. K., & Srivastava, M. Improved production of reducing sugars from rice straw using crude cellulase activated with Fe3O4/Alginate nano-composite. Bioresource technology. 2015; 183: 262-266.

[21] Ilyas, R. A., Sapuan, S. M., Ibrahim, R., Atikah, M. S. N., Atiqah, A., Ansari, M. N. M., & Norrrahim, M. N. F. Production, processes and modification of nanocrystalline cellulose from agro-waste: a review. Nanocrystalline materials. 2019: 3-32.

[22] Lakouraj, M. M., Mojerlou, F., & Zare, E. N. Nanogel and super-paramagnetic nanocomposite based on sodium alginate for sorption of heavy metal ions. Carbohydrate Polymers. 2014; 106: 34-41.

[23] Mittal, A., Ahmad, R., & Hasan, I. Poly (methyl methacrylate)-grafted alginate/Fe3O4 nanocomposite: synthesis and its application for the removal of heavy metal ions. Desalination and Water Treatment. 2016; 57(42): 19820-19833.

[24] Phiri, I., Ko, J. M., Mushonga, P., Kugara, J., Onani, M. O., Msamadya, S., ... & Madzvamuse, A. Simultaneous removal of cationic, anionic and organic pollutants in highly acidic water using magnetic nanocomposite alginate beads. Journal of Water Process Engineering. 2019; *31*: 100884.

[25] Isawi, H. (2020). Using zeolite/polyvinyl alcohol/sodium alginate nanocomposite beads for removal of some heavy metals from wastewater. Arabian Journal of Chemistry. 2020; 13(6): 5691-5716.

[26] Jayalakshmi, R., & Jeyanthi, J. Simultaneous removal of binary dye from textile effluent using cobalt ferrite-alginate nanocomposite: Performance and mechanism. Microchemical Journal. *2019*; *145*: 791-800.

[27] Yu, Y., Zhang, G., & Ye, L. Preparation and adsorption mechanism of polyvinyl alcohol/graphene oxide-sodium alginate nanocomposite hydrogel with high Pb (II) adsorption capacity. Journal of Applied Polymer Science. 2019; 136(14): 47318.

[28] Karkeh-Abadi, F., Saber-Samandari, S., & Saber-Samandari, S. The impact of functionalized CNT in the network of sodium alginate-based nanocomposite beads on the removal of Co (II) ions from aqueous solutions. Journal of hazardous materials. 2016; *312*: 224-233.

[29] Gokila, S., Gomathi, T., Sudha, P. N., & Anil, S. Removal of the heavy metal ion chromiuim (VI) using Chitosan and Alginate nanocomposites. International journal of biological macromolecules. 2017; 104: 1459-1468.

[30] Cataldo, S., Muratore, N., Orecchio, S., & Pettignano, A. Enhancement of adsorption ability of calcium alginate gel beads towards Pd (II) ion. A kinetic and equilibrium study on hybrid Laponite and Montmorillonite–alginate gel beads. Applied Clay Science. 2015; 118: 162-170.

[31] Esmat, M., Farghali, A. A., Khedr, M. H., & El-Sherbiny, I. M. (2017). Alginate-based nanocomposites for efficient removal of heavy metal ions. International journal of biological macromolecules, *102*, 272-283.

[32] Zhong, R., Yu, F., Schutyser, W., Liao, Y., de Clippel, F., Peng, L., & Sels, B. F. Acidic mesostructured silica-carbon nanocomposite catalysts for biofuels and chemicals synthesis from sugars in alcoholic solutions. Applied Catalysis B: Environmental. 2017; 206: 74-88.

[33] Budhi, S., Mukarakate, C., Iisa, K., Pylypenko, S., Ciesielski, P. N., Yung, M. M., ... & Trewyn, B. G. Molybdenum incorporated mesoporous silica catalyst for production of biofuels and value-added chemicals via catalytic fast pyrolysis. Green Chemistry. 2015; *17*(5): 3035-3046.

[34] Shen, Y. Rice husk silica derived nanomaterials for sustainable applications. Renewable and Sustainable Energy Reviews. 2017; 80: 453-466.

[35] Elimbinzi, E., Nyandoro, S. S., Mubofu, E. B., Manayil, J. C., Lee, A. F., & Wilson, K. Valorization of rice husk silica waste: organo-amine functionalized castor oil templated mesoporous silicas for biofuels synthesis. Microporous and Mesoporous Materials. 2020; 294: 109868.

[36] Anuar, M. F., Fen, Y. W., Zaid, M. H. M., Matori, K. A., & Khaidir, R. E. M. Synthesis and structural properties of coconut husk as potential silica source. Results in Physics. 2018; *11*: 1-4.

[37] Naqvi, J., Shah, F. H., & Mansha, M. Extraction of amorphous silica from wheat husk by using KMnO4. Journal of faculty of engineering & technology. 2011; 18(1): 39-46.

[38] Osman, N. S., & Sapawe, N. Waste material as an alternative source of silica precursor in silica nanoparticle synthesis–a review. Materials Today: Proceedings. 2019;19: 1267-1272.

[39] Raja, S. A Review: Sources of Silica from Agro Waste and Its Extraction Methods JARDCS. 2019; 11: 765-772.

[40] Sapawe, N., Surayah Osman, N., Zulkhairi Zakaria, M., Amirul Shahab Syed Mohamad Fikry, S., & Amir Mat Aris, M. Synthesis of green silica from agricultural waste by sol-gel method. Materials Today: Proceedings. 2018; 5(10): 21861-21866. DOI:10.1016/j.matpr.2018.07.043

[41] Salama, A., Diab, M. A., Abou-Zeid, R. E., Aljohani, H. A., & Shoueir, K. R. Crosslinked alginate/silica/zinc oxide nanocomposite: a sustainable material with antibacterial properties. Composites Communications. 2018; 7: 7-11.

[42] Yu, G., Lu, Y., Guo, J., Patel, M., Bafana, A., Wang, X., ... & Wujcik, E. K. Carbon nanotubes, graphene, and their derivatives for heavy metal removal. Advanced Composites and Hybrid Materials. 2018; 1(1): 56-78.

[43] Yadav, S., Asthana, A., Chakraborty, R., Jain, B., Singh, A. K., Carabineiro, S. A., ... & Hassan, A. B. Cationic dye removal using novel magnetic/activated charcoal/β-cyclodextrin/alginate polymer nanocomposite. Nanomaterials. 2020; 10(1): 170.

[44] Konwar, A., & Chowdhury, D. Property relationship of alginate and alginate–carbon dot nanocomposites with bivalent and trivalent cross-linker ions. RSC Advances. 2015; 5(77): 62864-62870.

[45] Ma, J., Jiang, Z., Cao, J., & Yu, F. Enhanced adsorption for the removal of antibiotics by carbon nanotubes/graphene oxide/sodium alginate triple-network nanocomposite hydrogels in aqueous solutions. Chemosphere. 2020; 242: 125188.

[46] Mohan, D., Singh, K. P., Sinha, S., & Gosh, D. Removal of pyridine

derivatives from aqueous solution by activated carbons developed from agricultural waste materials. Carbon. 2005; 43(8): 1680-1693.

[47] Kerdnawee, K., Termvidchakorn, C., Yaisanga, P., Pakchamsai, J., Chookiat, C., Eiad-Ua, A., ... & Charinpanitkul, T. Present advancement in production of carbon nanotubes and their derivatives from industrial waste with promising applications. KONA Powder and Particle Journal. 2017: 2017001.

[48] Somanathan, T., Prasad, K., Ostrikov, K. K., Saravanan, A., & Krishna, V. M. Graphene oxide synthesis from agro waste. Nanomaterials. 2015; 5(2): 826-834.

[49] Hesas, R. H., Daud, W. M. A. W., Sahu, J. N., & Arami-Niya, A. The effects of a microwave heating method on the production of activated carbon from agricultural waste: A review. Journal of Analytical and Applied pyrolysis. 2013; 100: 1-11.

[50] Ensuncho-Muñoz, A. E., & Carriazo, J. G. Characterization of the carbonaceous materials obtained from different agro-industrial wastes. Environmental technology. 2015; 36(5): 547-555.

Chapter 4

Alginate Metal Complexes and Their Application

Ghaidaa Alkhayer

Abstract

Alginate is a natural polymer that can form complexes in the presence of multivalent metal. In this chapter, we summarized the newest alginate metal complexes application in many fields; organic synthesis, environmental and medical application. The main idea was about alginate complexes' role in the drug delivery system as a chiral excipient to reach the enantioselective release in the case of chiral drugs. We also present a case study about the ketoprofen enantioselective release investigation from alginate mixed beads with two ion metal types.

Keywords: Alginate-metal complex, ionotropic method, chiral excipient, enantioselective release, enantiomers, chiral HPLC

1. Introduction

Many publications in the last decade dealt with different applications of alginate in several fields. Alginates application is depending on its source, extraction methods, it's physiological characteristics, functions, and properties [1]. Developing alginate and its derivatives were designed in various formulations for biomedical applications; such as wound dressing, tissue engineering, drug delivery, and dental application. Numerous natural polymers have been investigated for the development of different drug delivery systems [2–4]. For this use, alginate was developed and applied in drug delivery systems in form of capsules, hydrogels, tablets, nanoparticles, beads, microspheres, films, membranes, and others [4–7].

In the process, the chapter focus on the alginate metal complex preparation, application of the prepared complexes, a comparison of the release behavior between different alginate metal complex loaded with two chiral drugs (Profens). This includes the effect of bead kind on the enantioselective release (ESR) and the release mechanism due to the chiral interaction between alginate complexes and chiral drugs. Finally, the case study section discusses ketoprofen-loaded beads preparation in the presence of two ion metal types and an In-vitro ESR study for the prepared beads during the experiment time. Therefore, this chapter summarizes our current thought about alginate metal complex application as an ESR agent in addition to its role in many other fields.

2. General properties of alginate

Sodium alginate is the most common salt of alginic acid, it is a water-soluble and natural nontoxic polysaccharide extracted from marine brown algae. It contains 2

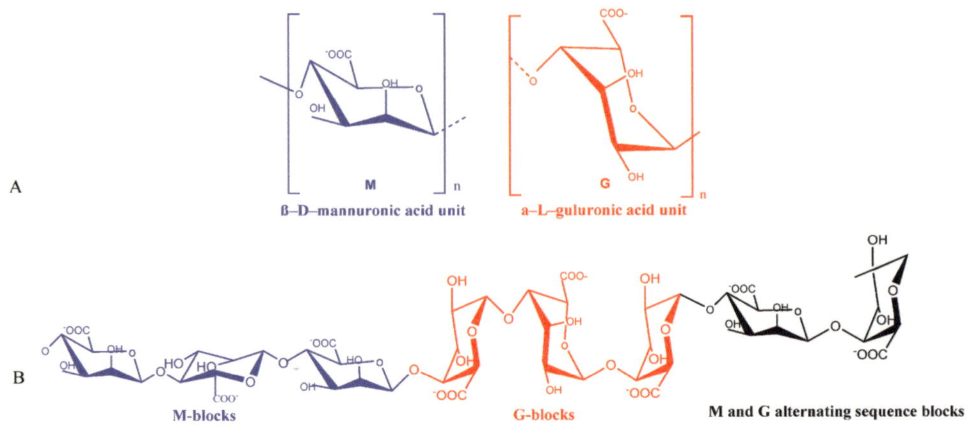

Figure 1.
Chemical structure of alginate, A: alginate monomers, B: structures of G-block, M-block, and alternating block in alginate.

uronic acids, β-D-mannuronic acid (M) and α-L-guluronic acid (G) [8], and it is composed of homopolymeric blocks MM or GG, and blocks with an alternating sequence (MG blocks) [9, 10], **Figure 1**. The alginate's rigidity decreases along with the series GG > MM > MG due to its different contents of M and G; which depends on alginates' different sources. On the other hand, the divalent metal ions affinities to alginate are dependent on the M: G units' ratio.

The alginate's affinity for divalent ions increases in the order [9]:

a. Alginate from *Laminaria digitata* rich with M units:

Pb > Cu > Cd > Ba > Sr > Ca > Co, Ni, Zn, Mn > Mg

b. Alginate from *Laminaria Hyperborea* rich with G units:

Pb > Cu > Ba > Sr > Cd > Ca > Co, Ni, Zn, Mn > Mg

The divalent cations concentration for complex formation from the two types of seaweeds is the same and follow the order:

Ba < Pb < Cu < Sr < Cd < Ca < Zn < Ni < Co < Mg.

3. Alginate metal complexes preparation by Ionic crosslinking

The ease of beads preparation and the mild conditions of alginate metal complexes preparation make it very unique compared to other polysaccharides. The ability to ion binding is selectively linked to the guluronate units (G). The M/G ratio, G-block length, and sequence of M and G blocks are the most important factors affecting the resulted alginate complexes.

Alginate forms hydrophilic gels by interaction with multivalent metal ions [9]. Since alginate gel can easily be formed by this ionic interaction in an aqueous medium; gel beads are commonly obtained by dropping solutions of sodium alginate into solutions of ion metal chloride [11–21], **Figure 2**. Generally, calcium chloride is one of the most commonly used as an ionically cross-linked agent

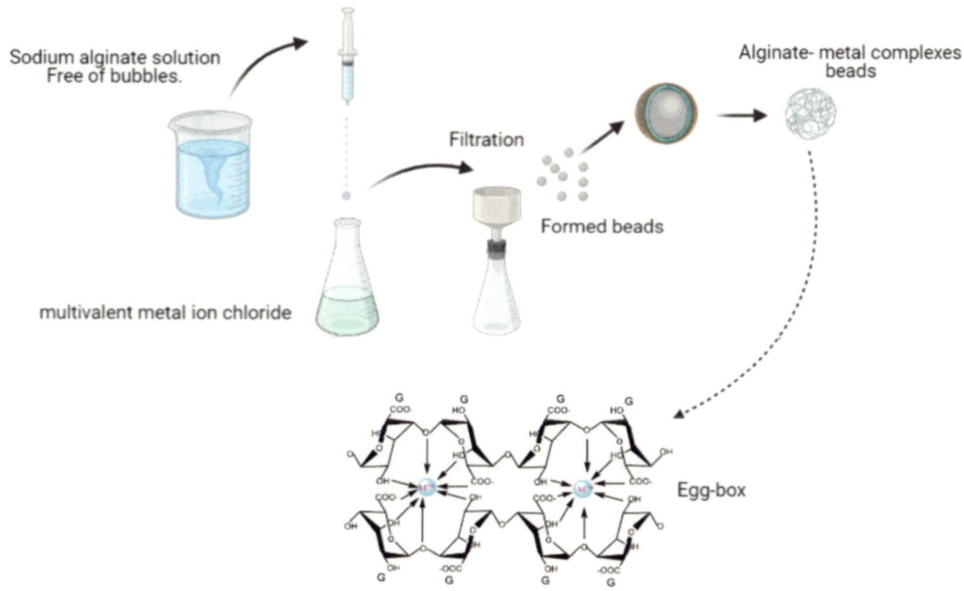

Figure 2.
Alginate gelation by ionic interaction between alginate and a multivalent cation.

to alginate. On the other hand, the gelation rate is depending on many factors, such as congealing time, temperature, and congealing sequence in the case of mixed beads (have two ion types). The formed egg-box has cavities between the G blockchains and the multivalent cation. In which the cavity fits the used cation [9].

4. Applications of alginate metal complexes

4.1 Organic synthesis field

Click chemistry is a new approach for drug industries based on the chemical reactions with high yields, and stereo-selectivity results with low reaction time. Bahsis et all [22] studied the synthesis of hydrogel catalyst; consist of sodium alginate with copper (II) for the azide-alkyne cycloaddition in the form of spherical beads. The prepared beads showed high catalytic activity for the required interaction.

Pua et al. [23] studied the synthesis of three alginate catalysts for the esterification of oleic acid. Ferric-alginate, Copper-alginate, and Nickel alginate beads were used to esterify the free fatty acid, and Fe- beads were the most successful ones.

In the work of Souza et al. [24], Alg-Cu^{+2} microspheres were prepared via an ion exchange process, and it was examined as a catalyst for the synthesis of some substituted pyrazoles. The resulted product was in excellent yield, and the catalyst activity still good even after five reactions.

Qiao et al. [25] introduced a new hybrid material of Ni-alginate beads. They distinguished with their remarkable activity and stability as styrene hydrogenation catalyst with recycling ability for 20 times. On the other hand, the ease of this hybrid material preparation allows examining its hydrogenation activity of unsaturated substrates.

4.2 Environmental field

In the past decades, many researchers have been aware of the heavy metals that affected the environment, due to Pollution caused by mining and different manufacturing. The industrial wastewaters clean-up of toxic metals such as Pb, Hg, Ni, Cr, Cd, As is challenging for many research centers [26–28].

Membrane filtration, electrodeposition, ion exchange, and chemical precipitation were the most techniques involved in removing metal from aqueous contaminated solutions. However, there were disadvantages to some of these methods compared to the treatment of complex [29]. Sodium alginate is one of the rawest materials using in synthesis methods as adsorbents to remove heavy metal ions from aqueous solutions.

Gao et al. [30] reviewed the Possibility of developing the sodium alginate as adsorbents, the involved mechanisms in the adsorption process were; electrostatic interaction, ion exchange, reduction, and photocatalytic reduction. In another study, they provide a synthesized sodium alginate adsorbent which showed notable selectivity towards Pd (II). Therefore, they provide selective industrial applications to reduce the Pd (II) from effluents.

Clinoptilolite/Nickel Ferrite/Sodium Alginate Nanocomposite beads were prepared by Bayat et all. [31] via many stages to remove methylene blue dye from water. Pseudo-second-order was the best fit model for adsorption kinetic, and the optimal pH was 5 for methylene blue adsorption.

4.3 Pharmaceutical field

The development of a drug delivery system is one of the most researcher's concerns. Particularly, in the case of chiral excipients [7, 32–34], which leads to possible steric interactions between chiral excipient and the chiral drug due to enantioselective release. Thus, it could affect the pharmacological and bioavailability studies of chiral drugs. Many chiral excipients were used in several pharmaceutical formulations, and numerous researchers have studied the effect of chirality on the drug release [35] such as ketoprofen [36, 37], propranolol [38, 39], metoprolol [40], tiaprofenic acid [41], ibuprofen [42], salbutamol [43, 44] and verapamil [45] from its formulations.

Sodium alginate can interact with multivalent metal ions leading to the proposed egg-box model [14]. Thus, drug-loaded beads could be prepared by the ionotropic gelation method, this allows the study of drug release behavior.

Alginate's common role in pharmaceutical industries includes gel-forming, stabilizing, and thickening agents. Nowadays, it can play an important role in drug-controlledrelease [8, 10]. The most frequent use of alginate and/or its derivatives is in oral dosage forms, but the use of alginate metal complex is still under investigation in many cases, especially in the case of studying the drug release behavior. Here, we briefly describe the use of the alginate metal complex in sustained and enantioselective release for some chiral drugs.

4.3.1 Sustained release applications

Alginates were classified among the most varied biopolymers, due to their flexibility for modification. Thus, it was widely used in food, drugs, and cosmetics. This kind of polymers could be useful as an excipient for sustained and controlled drug delivery. Therefore, many researchers introduced the use of alginate in the pharmaceutical field and biomedical applications. **Table 1** summarizes some examples of the alginate metal complex's application as sustained or prolonged drug release agents.

Type of complex	Dosage form	Drug	Remarks	Ref.
Calcium alginate with acrycoat E30D	Microparticles	Ketoprofen	The ketoprofen release from the prepared microparticles was significantly prolonged	[46]
Chitosan- calcium alginate with PNI-PAAM*	Beeds	Indomethacin	The prepared beads were used as a pH/temperature-sensitive drug delivery system, and they could be useful for the controlled release of bioactive agents.	[47]
Calcium alginate with PNI-PAM	Semi-IPN Beads	Indomethacin	The prepared beads have the potential to be used as a pH/ temperature-sensitive drug delivery system.	[48]
Calcium alginate	Beeds	Trimetazidine	Generally, beads were prepared in two methods; the drug content was higher in the sequential and simultaneous methods with increased $CaCl_2$ and polymer concentration, but lower with increased drug concentration in the sequential method.	[49]
Calcium alginate with HPMC	Microbeads	Flurbiprofen	Flurbiprofen was successfully loaded with high efficiency and prolonged release from the prepared beads.	[50]
Calcium alginate	Microspheres	Ketoprofen	The prepared microspheres could be used for sustaining drug release, and the increasing of polymer concentration leads to slower drug release.	[51]
Calcium alginate	Beeds	Propranolol	The drug content increased with decreasing Ca^{2+} concentration in the prepared beads.	[52]
Calcium alginate with chitosan-	Beeds	protein drugs	The formed beads could preserve the bioactivity of the studied drugs due to drug loading in an aqueous medium.	[53]
Calcium alginate	Beeds	5-fluorouracil	The prepared beads were designed to release the maximum drug release in the colon.	[54]
Calcium alginate	Beeds	Ketoprofen	The prepared beads showed interesting results for fast delivery to the upper gastrointestinal tract.	[55]
Carboxymethyl Chitosan with Calcium alginate	Beeds	Insulin	The pH-sensitive prepared beads were loaded with insulin at different weight ratios, and the released insulin was stable and biologically active. The beads could be useful for insulin as an oral dosage form.	[56]

Type of complex	Dosage form	Drug	Remarks	Ref.
Zink alginate	Beeds	Ketoprofen	Zn-alginate beads were prepared and loaded with ketoprofen. In vitro and in vivo release were studied, and the result showed the beads could be suitable for a delayed release of anti-inflammatory drugs.	[57]
Calcium alginate	Microspheres	Risperidone	The microspheres showed sustained drug release, and they were feasible to serve the delivery system of therapeutic agents.	[58]
Calcium alginate with carbopol	Microbeads	Clarithromycin	The release of clarithromycin from microbeads showed promising results in vitro for the eradication of *H. pylori* infection.	[59]
Calcium alginate with chitosan	Beeds	Ketoprofen	The chitosan-alginate beads showed a sufficient sustained release of ketoprofen and low gastrointestinal irritation.	[60]

PNI-PAAM: poly(N-isopropylacrylamide), Semi-IPN: semi-interpenetrating, HPMC: Hydroxyl propyl methyl cellulose.

Table 1.
Some examples of alginate metal complex application as sustained release agent.

4.3.2 Enantioselective release applications

Academic researchers recognized the importance of developing chiral drugs and their pharmaceutical industry. Investigation of enantioselective release (ESR) was discussed with two main strategies: 1- chiral interactions between a chiral drug and chiral matrices, 2- key-to-lock strategy with molecular-imprinting polymers [35, 61]. Several publications discussed the ESR, but few of them dealt with the alginate metal complex as a chiral excipient. This review focuses on the alginate complexes and their role in some profens ESR.

As mentioned above, alginate metal complexes have been extensively used in the pharmaceutical field. However, the enantioselective release of chiral drugs from alginate complexes is very rare.

Our previous studies were among the first publications in this field [62–64]. Alginate metal complexes in form of beads were prepared by the ionotropic gelation method. Ketoprofen (KTP) was loaded in the first group of beads [62], and tiaprofenic acid (Tia) was loaded in the second one [63]. In all cases, the resulted beads were characterized; bead size, metal content, shrinkage ratio, drug loading, and loading efficiency were calculated. The in-vitro release was carried out in an aqueous phosphate buffer that resembles gastric medium (6.8–7.4), and the enantioselective release (ESR) was observed in many complexes [62–64].

Beads in the two groups tend to have a metal content higher than the calculated ones. These results may due to the retention of free ions in the resulted network. On the other hand, in both cases, the divalent ion metal beads show a smaller size than the beads with trivalent ones. **Figure 3** shows the drug-loaded beads (KTP and Tia) metal contents compared to blank beads.

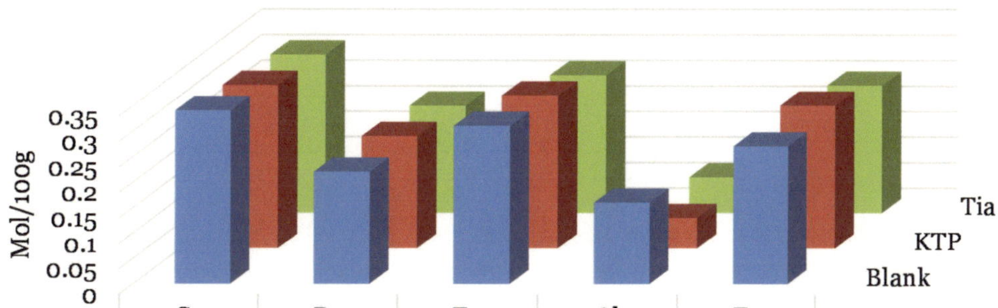

	Ca	Ba	Zn	Al	Fe
Blank	0.34	0.22	0.31	0.16	0.27
KTP	0.32	0.22	0.3	0.06	0.28
Tia	0.31	0.21	0.27	0.07	0.25

Figure 3.
Comparison of metal content (mol/100 g) for blank beads (blue column), KTP loaded beads (red column), and Tia loaded beads (green column).

In order to explore the ESR result, the IR spectrum for all prepared beads types was determined at a range of 4000–400 cm^{-1} [62, 63]. There was an obvious hydrogen bonding between hydroxyl and carboxyl groups of alginates with the Tia and KTP ketone and carboxylic hydroxyl. The OH signals of KTP and Tia and the alginates' OH combines together in one signal due to hydrogen bonding interaction which could explain the ESR results. More discussion was described in detail in references [62, 63].

ESR comparison between KTP and Tia loaded beads shows a similar ESR behavior for AZnK and AZnT beads as shown in **Figure 4**. In both cases, the ESR > 1 indicating to a stronger interaction with S- enantiomer meaning more

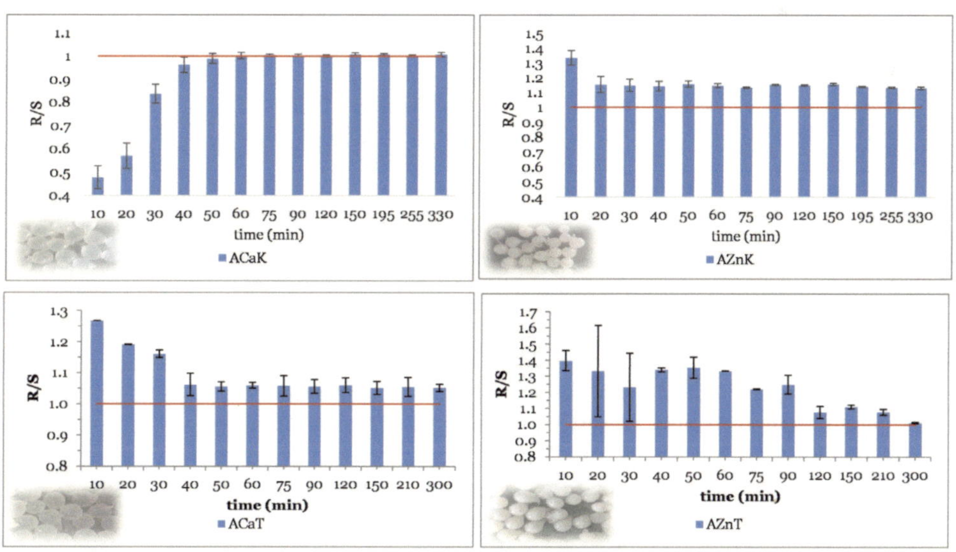

Figure 4.
ESR for KTP and Tia from divalent alginate-metal complexes beads as R/S ratio (blue column), racemic release R/S = 1 (red line).

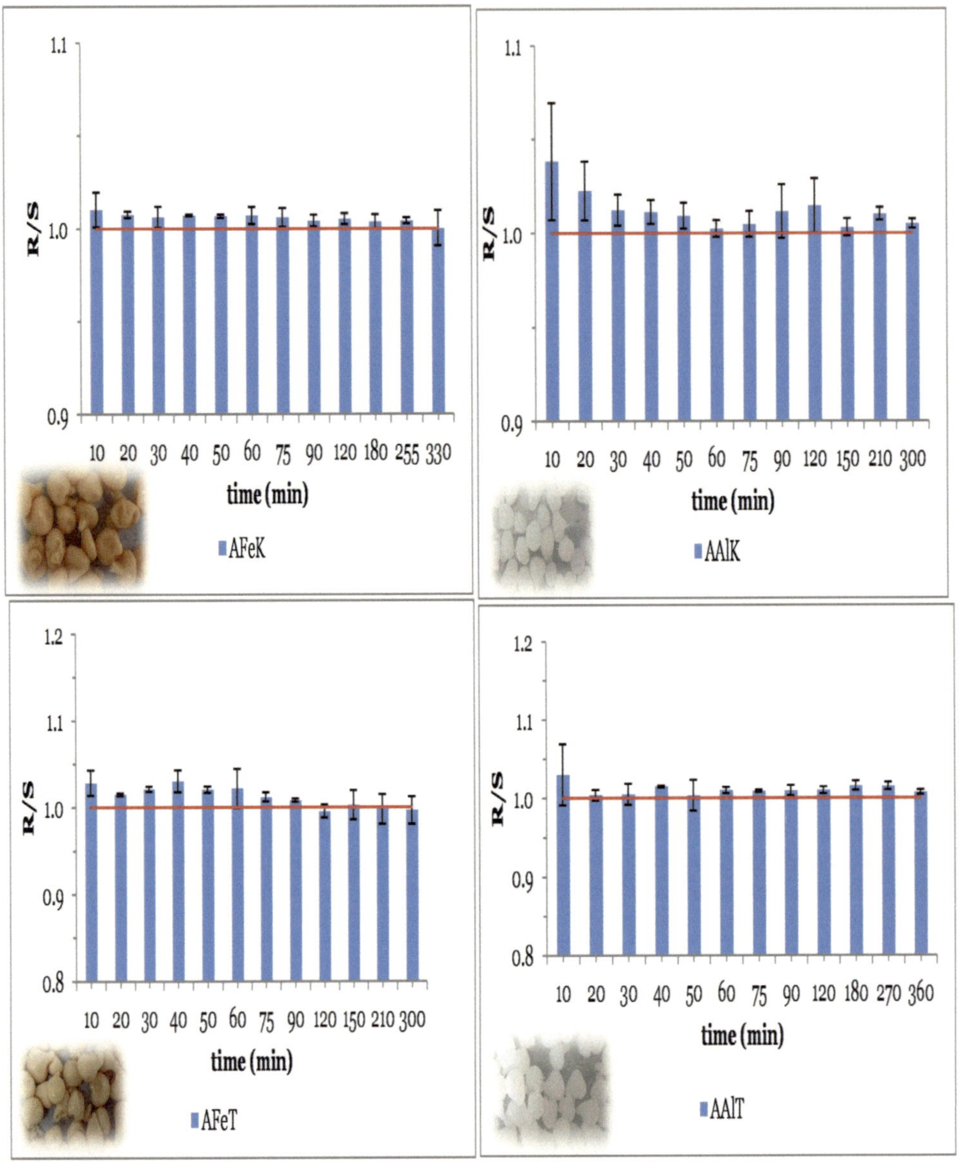

Figure 5.
ESR for KTP and Tia from trivalent alginate-metal complexes beads as R/S ratio (blue column), racemic release R/S = 1 (red line).

retention of S-enantiomer in contrast to R-enantiomer. **Figure 5** shows that there were not any significant differences between the trivalent loaded beads. In both cases, The ESR was almost =1 with the racemic release. However, ACa beads show an obvious ESR for both KTP and Tia but in a contrasting way. The ESR < 1 in ACaK indicating to strong chiral interaction with R- enantiomer. While ESR > 1 in ACaT indicating to strong chiral interaction with S- enantiomer. These different results in many cases due to the difference in KTP and Tia structures **Figure 6**.

On the other hand, the kinetic simulation of studied beads [63, 64] shows that the best fit models for each enantiomer and the racemic mixture were the same. However, the obtained models differ depending on the type of complexation due to the resulting "egg-box" structure.

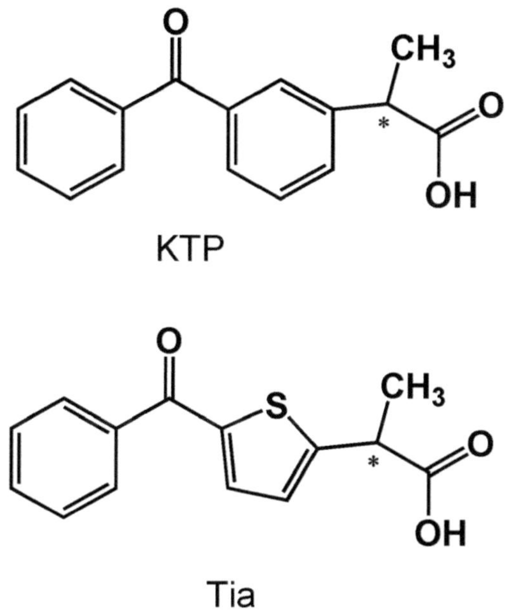

Figure 6.
Tiaprofenic acid (Tia) and ketoprofen (KTP) chemical structure, asymmetric carbon labeled with.*

5. Case study

Case study 1: Enantioselective release of ketoprofen enantiomers form alginate complexes with two ion metal type.

5.1 Drug loading and beads preparation

An ionotropic method was used to prepare alginate metal complex beads. Racemic ketoprofen was dissolved with sodium alginate in phosphate buffer solution (PBS, pH = 7.4). The final ratio was 1 to 3.75 (w/w) [62–64]. The congealing solution contained a metal chloride (3% w/v) at room temperature. Beads were commonly formed by dropping the alginate solution into the metal chloride. In this case, beads congealed with barium chloride for 3 hours. After that, the congealing continued with the other metal chloride (Ca, Zn, Al, Fe), and the total congealing time 24 hours. Beads were separated and dried at 40°C for 48 h, the method was described in detail in Refs. [62–64]. However, beads were released in PBS with pH = 7.4. 0.2 ml of aliquot was taken at different time intervals, and replenished to the release medium with fresh PBS to maintain the volume to 5 ml. the aliquot was extracted and analyzed by chiral HPLC.

Drug loading (KTP%) and loading efficiency (L%) were determined by the following equations:

$$\text{Drug loading}\left(\text{KTP\%}\right) = \left(W_l / W_{db}\right) * 100 \tag{1}$$

$$\text{Loading efficiency}\left(\text{L\%}\right) = \left(W_l / W_t\right) * 100 \tag{2}$$

Where W_l: loaded KTP weight, W_t: Initial KTP weight, and W_{db}: total dried beads weight. The shrinkage ratio was calculated using the formula:

Beads name	Initial weight		Formed beads				Dried beads		S%
	W_{Alg} (mg)	KTP W_t (mg)	Wet beads (g)	KTP W_{res} (mg)	KTP W_l (mg)	KTP L%	W_{db} (mg)	KTP %	
ABaCaK	316	83	8.490	11.21	71.8	86.5	469	15.3	57.8
ABaZnK	315	81	8.286	7.86	73.1	90.2	495	14.8	60.9
ABaAlK	312	80	10.544	3.75	76.3	95.4	558	13.7	64.7
ABaFeK	317	81	10.282	1.63	79.4	80.0	532	14.9	62.2

W_{Alg}: Initial alginate weight, W_{res}: resediual KTP in congealing bath.

Table 2.
Drug loading and loading efficiency results for of the prepared beads.

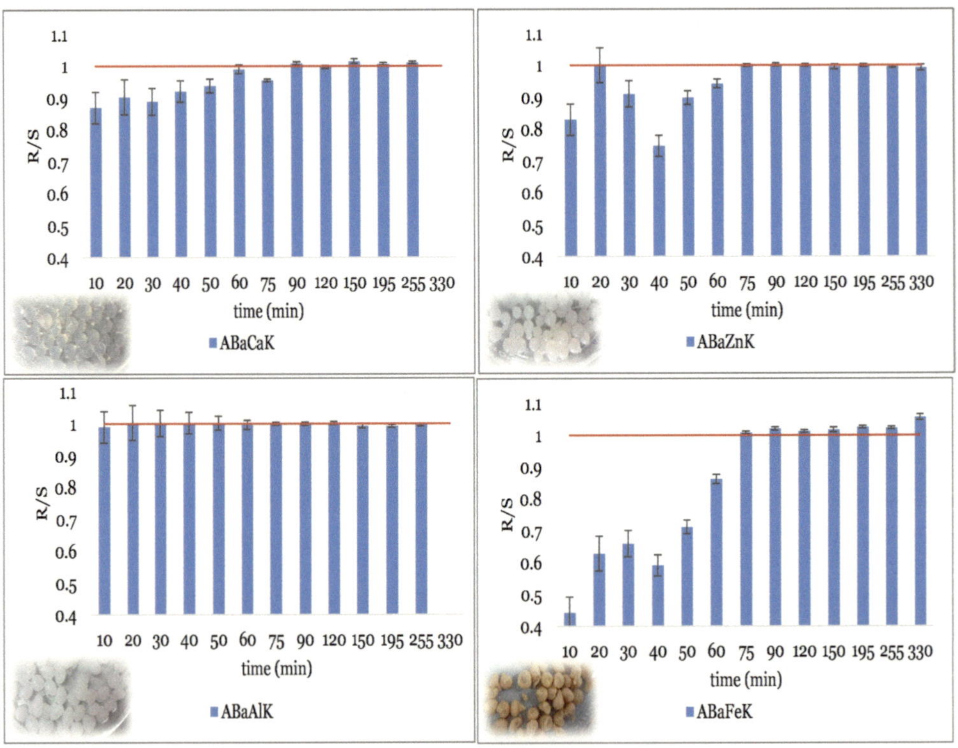

Figure 7.
ESR for KTP from mixed alginate-metal complexes beads as R/S ratio (blue column), racemic release R/S = 1 (red line).

$$\text{Shrinkage ratio}\,(S\%) = \left(\left(D_b - D_a\right)/D_b\right) * 100 \qquad (3)$$

Where, D_b and D_a are the diameters of the beads before and after drying respectively.

Table 2 shows that KTP% values varied from 13.7 for ABaAlK to 15.3 for ABaCaK, while ABaAlK and ABaFeK have the highest shrinkage ratio due to the trivalent ion metal in the formed complex.

5.2 Enantioselective release study

Chiral HPLC was used to monitor the ESR and expressed as the R/S enantiomers ratio of chromatographic area. The mobile phase consists of hexane: isopropanol: TFA (90:10:0.1 v/v%), with a 1 ml min^{-1} flow rate. The column, Kromasil®-5-amy- coat (250 X 4.6 mm i.d) 5 mm, was equilibrated for at least 30 min, at temperature (30 C°). ESR values in **Figure 7** shows that ESR <1 for ABaCaK, ABaZnK, and ABaFeK beads within the first 60 min due to a strong interaction with R- enantiomer; which retains in the beads for more time compared to S- enantiomer. However, ESR had opposite behaviors; ESR < 1 for ABaZnK compared with AZnK result in **Figure 4**, while ABaFeK shows an obvious ESR comparing to AFeK in **Figure 5** indicating the role of the mixed congealing with Ba to alter the KTP release behavior. In fact, no significant ESR was obtained for ABaAlK and the release was practically racemic all over the experiment time. These results suggest differences in egg-box stereochemistry due to different binding strengths depending on the congealing method, ion metal type, and the complexation kind (with one or mixed metals).

6. Conclusions

Natural and biodegradable polymers were used increasingly in pharmaceutical formulations, food, and some industrial applications. This chapter has introduced various aspects of alginate metal complexes preparation and its application in organic synthesis, environmental and pharmaceutical fields as a chiral excipient. The latest case involved alginate's ability to build complexes in the presence of multivalent metal. The prepared beads were loaded with racemic profens by an ionotropic method, and the chiral interactions are assumed to affect the drug release due to alginates and profen's chirality, by an in-vitro release in the aqueous solution resembles an intestine medium (6.8–7.4), and the enantioselective release (ESR) was observed in many complexes and differ depending on the alginate-metal complexes type.

Acknowledgements

The author thanks the Research Institute for Pharmaceutical and Chemical Industries and the Higher Institute for Applied Sciences and Technology for providing the assistance needed to complete the work.

Abbreviations

ESR	enantioselective release
KTP	Ketoprofen
Tia	Tiaprofenic acid
Alg	alginate
ACaK	alginate calcium beads loaded with KTP
ACaT	alginate calcium beads loaded with Tia
AZnK	alginate zink beads loaded with KTP
AZnT	alginate zink beads loaded with Tia

AFeK	alginate iron (III) beads loaded with KTP
AFeT	alginate iron (III) beads loaded with Tia
AAlK	alginate aluminum beads loaded with KTP
AAlT	alginate aluminum beads loaded with Tia
PBS	phosphate buffer solution

Author details

Ghaidaa Alkhayer[1,2]

1 Research Institute for Pharmaceutical and Chemical Industries, Damascus, Syria

2 Higher Institute for Applied Sciences and Technology, Damascus, Syria

*Address all correspondence to: ghaidaa.s.alkhayer@gmail.com

References

[1] Puscaselu RG, Lobiuc A, Dimian M, et al. Alginate: From food industry to biomedical applications and management of metabolic disorders. *Polymers (Basel)* 2020; 12: 1-30.

[2] Shin DY, Park JU, Choi MH, et al. Polydeoxyribonucleotide-delivering therapeutic hydrogel for diabetic wound healing. *Sci Rep* 2020; 10: 1-15.

[3] Kamaly N, Yameen B, Wu J, et al. Degradable controlled-release polymers and polymeric nanoparticles: Mechanisms of controlling drug release. *Chem Rev* 2016; 116: 2602-2663.

[4] Mantha S, Pillai S, Khayambashi P, et al. Smart Hydrogels in Tissue Engineering and. *Materials (Basel)* 2019; 12: 33.

[5] Mørch ÝA. *Novel alginate microcapsules for cell therapy*. 2008.

[6] Hasnain MS, Nayak AK. *Alginate-inorganic composite particles as sustained drug delivery matrices*. Elsevier Inc. Epub ahead of print 2018. DOI: 10.1016/B978-0-12-813741-3.00003-0.

[7] Abasalizadeh F, Moghaddam SV, Alizadeh E, et al. Alginate-based hydrogels as drug delivery vehicles in cancer treatment and their applications in wound dressing and 3D bioprinting. *J Biol Eng* 2020; 14: 1-22.

[8] Ahmad Raus R, Wan Nawawi WMF, Nasaruddin RR. Alginate and alginate composites for biomedical applications. *Asian J Pharm Sci* 2021; 1-27.

[9] Hasnain MS, Nayak AK. *Alginates Versatile, Versatile Polymers in Biomedical Applications and Therapeutics*. 2019. Epub ahead of print 2019. DOI: 10.1017/CBO9781107415324.004.

[10] Lee KY, Mooney DJ. Alginate: Properties and biomedical applications. *Prog Polym Sci* 2012; 37: 106-126.

[11] Vicini S, Mauri M, Wichert J, et al. Alginate Gelling Process: Use of Bivalent Ions Rich Microspheres. *Polym Eng Sci* 2017; 1-6.

[12] Sun D, Liu Y, Wu H, et al. *Effects of gelling bath on the physical properties of alginate gel beads and the biological characteristics of entrapped HepG2 cells*. 2018. Epub ahead of print 2018. DOI: 10.1002/bab.1585.

[13] Cheryl-low YL, Theam KL, Lee H V. Alginate-derived solid acid catalyst for esterification of low-cost palm fatty acid distillate. *ENERGY Convers Manag* 2015; 106: 932-940.

[14] Mørch ÄA, Donati I, Strand BL, et al. Effect of Ca 2+ , Ba 2+ , and Sr 2+ on Alginate Microbeads. *Society* 2006; 1471-1480.

[15] Mørch YA, Qi M, Gundersen POM, et al. Binding and leakage of barium in alginate microbeads. *J Biomed Mater Res - Part A* 2012; 100 A: 2939-2947.

[16] Yildiz A, Sensoy D, Baytokova A, et al. Preparation of sustained-release Tiaprofenic acid beads. *J Fac Plzarnz lstanblll* 2001; 34: 27-33.

[17] Kragovi, Milan, Snežana Pašali'c, Marija Markovi'c, Marija Petrovi'c, Blagoje Nedeljkovi'c MM and MS. Natural and Modified Zeolite—Alginate Composites. Application for Removal of Heavy Metal Cations from Contaminated Water Solutions. *Minerals* 2018; 8: 1-16.

[18] Wang N, Liu Q, Li Y, et al. Self-crosslink assisted synthesis of 3D porous branch-like Fe3O4/C hybrids for high-performance lithium/sodium-ion batteries. *RSC Adv* 2017; 7: 50307-50316.

[19] Horniblow RD, Latunde-dada GO, Harding SE, et al. The chelation of colonic luminal iron by a unique sodium

alginate for the improvement of gastrointestinal health. *Mol Nutr Food Res* 2016; 1-11.

[20] Konwar A, Chowdhury D. Property relationship of alginate and alginate-carbon dot nanocomposites with bivalent and trivalent cross-linker ions. *RSC Adv* 2015; 5: 62864-62870.

[21] Wan J, Tao T, Zhang Y, et al. Phosphate adsorption on novel hydrogel beads with interpenetrating network (IPN) structure in aqueous solutions: Kinetics, isotherms and regeneration. *RSC Adv* 2016; 6: 23233-23241.

[22] Bahsis L, Ablouh EH, Anane H, et al. Cu(ii)-alginate-based superporous hydrogel catalyst for click chemistry azide-alkyne cycloaddition type reactions in water. *RSC Adv* 2020; 10: 32821-32832.

[23] Pua FL, Looi KT, Shamala Gowri K, et al. Synthesis and characterization of different transition metal-alginate based heterogeneous catalyst for esterification reaction. *Key Eng Mater* 2016; 709: 57-60.

[24] Souza JF, De Aquino TFB, Nascimento JER. Alginate-copper microspheres as efficient and reusable heterogeneous catalysts for the one-pot synthesis of 4-organylselanyl-1: H-pyrazoles. *Catal Sci Technol* 2020; 10: 3918-3930.

[25] Qiao X, Niu L, Zhang H, et al. Controllable fabrication of a novel porous Ni-alginate hybrid material for hydrogenation. *Appl Catal B Environ* 2017; 218: 721-730.

[26] Muedi VM and KL. Environmental Contamination by Heavy Metals. In: *Intech*, pp. 115-133.

[27] Motlochová M, Slovák V, Pližingrová E, et al. Highly-efficient removal of Pb(II), Cu(II) and Cd(II) from water by novel lithium, sodium and potassium titanate reusable microrods. *RSC Adv* 2020; 10: 3694-3704.

[28] de Oliveira Sousa Neto V, Saraiva GD, Carvalho TV, et al. *Heavy metal removal from industrial effluents using biopolymer membranes*. Elsevier Inc. Epub ahead of print 2020. DOI: 10.1016/b978-0-12-818134-8.00014-6.

[29] Gao X, Guo C, Hao J, et al. International Journal of Biological Macromolecules Adsorption of heavy metal ions by sodium alginate based adsorbent-a review and new perspectives. *Int J Biol Macromol* 2020; 164: 4423-4434.

[30] Gao X, Guo C, Hao J, et al. Selective adsorption of Pd (II) by ion-imprinted porous alginate beads: Experimental and density functional theory study. *Int J Biol Macromol* 2020; 157: 401-413.

[31] Bayat M, Javanbakht V, Esmaili J. Synthesis of zeolite/nickel ferrite/ sodium alginate bionanocomposite via a co-precipitation technique for efficient removal of water-soluble methylene blue dye. *Int J Biol Macromol* 2018; 116: 607-619.

[32] Rowe RC, Sheskey PJ, Quinn ME. *Handbook of Pharmaceutical Excipients*. Sixth edit. USA: Pharmaceutical Press and American Pharmacists Association, 2009.

[33] Christianah Adeyeye M. Use of Chiral Excipients in Formulations Containing Chiral Drugs. In: *Chirality in Drug Design and Development*. 2004. Epub ahead of print 2004. DOI: 10.1201/9780203021811.ch2.

[34] REDDY IK, MEHVA R. *Chirality in Drug Design and Development*. USA, 2004.

[35] Yu H, Yong X, Liang J, et al. Materials Established for Enantioselective Release of Chiral

Compounds. *Ind Eng Chem Res* 2016; 55: 6037-6048.

[36] Solinís MA, De La Cruz Y, Hernández RM, et al. Release of ketoprofen enantiomers from HPMC K100M matrices - Diffusion studies. *Int J Pharm* 2002; 239: 61-68.

[37] Álvarez C, Torrado JJ, Cadórniga R. Stereoselective drug release from ketoprofen and ricobendazole matrix tablets. *Chirality* 1999; 11: 611-615.

[38] Bodhibukkana C, Srichana T, Kaewnopparat S, et al. Composite membrane of bacterially-derived cellulose and molecularly imprinted polymer for use as a transdermal enantioselective controlled-release system of racemic propranolol. *J Control Release* 2006; 113: 43-56.

[39] Jantarat C, Tangthong N, Songkro S, et al. S-Propranolol imprinted polymer nanoparticle-on-microsphere composite porous cellulose membrane for the enantioselectively controlled delivery of racemic propranolol. *Int J Pharm* 2008; 349: 212-225.

[40] Guo Z, Du Y, Liu X, et al. Enantioselectively controlled release of chiral drug (metoprolol) using chiral mesoporous silica materials. *Nanotechnology*; 21. Epub ahead of print 2010. DOI: 10.1088/0957-4484/21/16/165103.

[41] Vakily M, Jama F. Human pharmacokinetics of Tiaprofenic Acid after Regular and Sustained Release Formulations: Lack of Chiral Inversion and Stereoselective Release. *J Pharm Sci* 1994; 83: 495-498.

[42] Cheng R, Liu J, Xie P, et al. Chiral, pH-sensitive polyacrylamide hydrogels: Preparation and enantio-differentiating release ability. *Polymer (Guildf)* 2015; 68: 246-252.

[43] Solinís MA, Lugará S, Calvo B, et al. Release of salbutamol sulfate enantiomers from hydroxypropy lmethylcellulose matrices. *Int J Pharm* 1998; 161: 37-43.

[44] Srichana T, Suedee R. Evaluation of stereoselective dissolution of racemic salbutamol matrices prepared with commonly used excipients and 1H-NMR study. *Drug Dev Ind Pharm* 2001; 27: 457-464.

[45] Aubry A -F, Wainer IW. An in vitro study of the stereoselective dissolution of (rac)-verapamil from two sustained release formulations. *Chirality* 1993; 5: 84-90.

[46] Rajesh KS, Khanrah A, Biswanath S. Release of Ketoprofen from alginate microparticles containing film forming polymers. *J Sci Ind Res (India)* 2003; 62: 985-989.

[47] Shi J, Alves NM, Mano JF. Chitosan coated alginate beads containing poly(N-isopropylacrylamide) for dual-stimuli-responsive drug release. *J Biomed Mater Res - Part B Appl Biomater* 2008; 84: 595-603.

[48] Shi J, Alves NM, Mano JF. Drug release of pH/temperature-responsive calcium alginate/poly(N-isopropylacrylamide) semi-IPN beads. *Macromol Biosci* 2006; 6: 358-363.

[49] Mandal S, Senthil Kumar S, Krishnamoorthy B, et al. Development and evaluation of calcium alginate beads prepared by sequential and simultaneous methods. *Brazilian J Pharm Sci* 2010; 46: 785-793.

[50] Badarinath A V., Ravi Kumar Reddy J, Mallikarjuna Rao K, et al. Formulation and characterization of alginate microbeads of flurbiprofen by ionotropic gelation technique. *Int J ChemTech Res* 2010; 2: 361-367.

[51] Al-tahami K. Preparation, characterization, and in vitro release of ketoprofen loaded alginate

microspheres. *Int J Appl Pharm ISSN-* 2014; 6: 9-12.

[52] Lim LY, Wan LSC. Propranolol hydrochloride binding in calcium alginate beads. *Drug Dev Ind Pharm* 1997; 23: 973-980.

[53] Lin YH, Liang HF, Chung CK, et al. Physically crosslinked alginate/N,O-carboxymethyl chitosan hydrogels with calcium for oral delivery of protein drugs. *Biomaterials* 2005; 26: 2105-2113.

[54] Patel H, Nagle A, Murthy R. Characterization of calcium alginate beads of 5-fluorouracil for colon delivery. *Asian J Pharm* 2008; 2: 241.

[55] Psimadas D, Georgoulias P, Valotassiou V, et al. Design of Alginate-Based Aerogel for Nonsteroidal Anti-Inflammatory Drugs Controlled Delivery Systems Using Prilling and Supercritical-Assisted Drying. *J Pharm Sci* 2012; 102: 185-194.

[56] Mukhopadhyay P, Sarkar K, Soam S, et al. Formulation of pH-responsive carboxymethyl chitosan and alginate beads for the oral delivery of insulin. *J Appl Polym Sci* 2013; 129: 835-845.

[57] Cerciello A, Auriemma G, Morello S, et al. Design and in Vivo Anti-Inflammatory Effect of Ketoprofen Delayed Delivery Systems. *J Pharm Sci* 2015; 104: 3451-3458.

[58] Al-Tahami K. Preparation of Alginate Microspheres for the Delivery of Risperidone. *Yemeni J Med Sci* 2014; 8: 5-5.

[59] Pal R, Bhadoria APS, Ramteke S. Preparation and characterization of sodium alginate-carbopol-934P based mucoadhesive microbeads. *Der Pharm Lett* 2011; 3: 1-11.

[60] Cheng B, Li D, Huo Q, et al. Two kinds of ketoprofen enteric gel beads (CA and CS-SA) using biopolymer alginate. *Asian J Pharm Sci* 2018; 13: 120-130.

[61] Song L, Pan M, Zhao R, et al. Recent advances, challenges and perspectives in enantioselective release. *J Control Release* 2020; 324: 156-171.

[62] Alkhayer G, Khudr H, Koudsi Y. Enantioselective Release Behavior of Ketoprofen Enantiomers from Alginate-metal Complexes, Monitored by Chiral HPLC. *Anal Bioanal Chem reserch* 2020; 7: 61-76.

[63] Alkhayer G, Khudr H, Koudsi Y. Spectroscopic and chromatographic investigation of chiral interactions between tiaprofenic acid and alginate–metal-complexes. *RSC Adv* 2020; 10: 35121-35130.

[64] Alkhayer G, Khudr H, Koudsi Y. In vitro kinetic release study of ketoprofen enantiomers from alginate metal complexes. *Futur J Pharm Sci* 2021; 7: 1-11.

Chapter 5

Algal Alginate in Biotechnology: Biosynthesis and Applications

Cagla Yarkent, Bahar Aslanbay Guler, Ceren Gurlek, Yaprak Sahin, Ayse Kose, Suphi S. Oncel and Esra Imamoglu

Abstract

Algae are recognized as the main producer of commercial alginate. Alginate produced using algae is located in the walls and intracellular regions of their cells. Its properties vary depending on the species, growing and harvesting seasons, and extraction methods. Alginate has attracted the attention of several industries, thanks to its unique properties such as its biodegradability, biocompatibility, renewability and lack of toxicity features. For example, it is considered a good encapsulation agent due to the transparent nature of the alginate matrices. Also, this biopolymer is recognized as a functional food in the food industry. It can be tolerated easily in human body and has the ability to reduce the risk of chronic diseases. Besides, it is used as an abrasive agent, antioxidant, and thickening and stabilizing agents in cosmetic and pharmaceutic industries. Generally, it is used in emulsion systems and wound dressing patches. Furthermore, this polysaccharide has the potential to be used in green nanotechnologies as a drug delivery vehicle via cell microencapsulation. Moreover, it is suitable to adopt as a coagulant due to its wide range of flocculation dose and high shear stability. In this chapter, the mentioned usage areas of algal alginate are explained in more detail.

Keywords: algae, algal alginate, immobilization, food, cosmetic, pharmaceutic, green nanotechnology, wastewater treatment

1. Introduction

Algae are photosynthetic eukaryotic organisms that are found in many environments such as sea, freshwater, and land and they are significantly important for oxygen production all around the world. Most of them are microscopic organisms, and their cell size can vary from 1 μm up to 10 m. There are around 72,500 algal species that produce different metabolites and products such as carbohydrates, proteins, vitamins, and many other secondary metabolites that have different benefits to humans and other organisms [1]. Since algae are exposed to stress in their nature, such as high UV radiation, salinity, desiccation and so on, their metabolites can have high antioxidant and anti-inflammatory activity, which make them valuable. They support almost all life forms in the biosphere, being a food source with high protein content (~20%) [2].

Alginate is an unbranched polymer which consists of two different residues; α-L-guluronic acid (G block) and β-d-mannuronic acid (M block) that are linearly

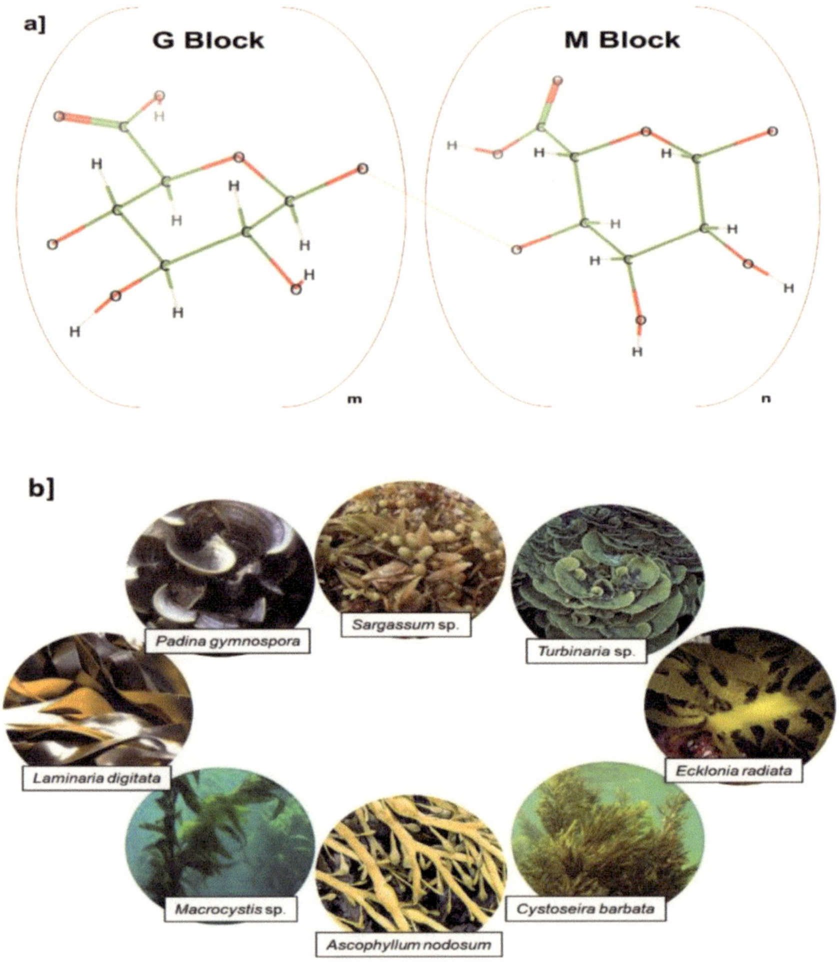

Figure 1.
(a) Chemical structure of G and M blocks; (b) Most used algae strains in alginate production [3–10].

linked together by 1–4 linkages to form the polymer as shown in **Figure 1a** [11, 12]. It is the most abundant biopolymer in the world and one of the primary carbohydrates in brown seaweeds [*Laminaria* sp., *Macrocystis* sp., *Lessonia* sp., etc. (**Figure 1b**)], reaching up to 40% of the dry weight depending on species [11–13]. A major source of alginate is the cell walls of brown seaweeds and their intracellular spaces [12]. Alginates are used commercially as thickening agents by the food and pharmaceutical industries as binders, gelling agents, and wound absorbents [11, 13]. Alginate and their derivatives and other forms such as zinc alginate, copper alginate, sodium calcium alginate, propylene glycol alginate, alginic acid, ester of alginic acid, and calcium, ammonium, and potassium salts are used in different industries in mostly textile industry with 50%, food industry follows it with 30%, and medical, cosmetic, and pharmaceutical industry with 20% of the annual production of 38,500 t alginate worldwide [12, 14].

The present book chapter focusses on alginate biosynthesis in algae and its extraction, immobilization of algae in alginate, and utilization of alginate in food and cosmetics sectors, pharmaceutical and biomedical applications, green nanotechnologies, and wastewater treatment as a coagulant.

2. Alginate biosynthesis in algae

For many years, alginate has attracted great interest in food, cosmetic, biomedical and pharmaceutical industries, and therefore the potential sources of alginate have been extensively studied to meet the commercial demand. Brown seaweeds also known as the marine macroalgae are recognized as the main producer of commercial alginate. These seaweeds (class *Phaeophyceae*) are algal species comprising complex multicellular brown algae with a wide range of sizes and morphologies [15, 16]. Their cell wall has a unique structure that contains phenolic compounds, proteins and high amount of carbohydrates. Among these components, alginate is the fundamental polysaccharide, which is found in the form of insoluble mixed salts of calcium, magnesium, sodium, barium, and potassium. There is also a large amount of alginate located in the intercellular matrix of algae and thus, total alginate content of biomass reaches up to 40% of dry weight [17]. The composition and the characteristics of alginate depend on the type of species, growth conditions, harvesting season, and extraction methods. In the work of Li et al. [18] it was shown that the chemical composition of the *Sargassum fusiforme* strain extensively varied during harvest and the highest alginate content was observed in June, whereas the alginate with maximum molecular weight and viscosity was obtained in May. In another study, a brown macroalgae *Treptacantha barbata* was cultured under four colors of light-emitting diode (LED) light including blue, red, green, and yellow and the blue LED light produced the highest sodium alginate content [19].

Today, all commercial alginates are obtained from algal sources and their composition varies among the species. The main genera containing a high amount of alginate are *Laminaria*, *Sargassum*, *Macrocystis*, *Ascophyllum*, *Lessonia*, *Ecklonia* and *Alaria* [20, 21]. Different macroalgae species and their alginate compositions are summarized in **Table 1**.

Previous metabolic studies have focused on the investigation of biological pathway of alginate synthesis in brown algae and bacteria that is another source of alginate. Despite the advances in molecular biology and genomic studies, the biosynthesis pathway and regulatory mechanism of alginate in algae have been poorly characterized. However, several studies of bacterial and algal alginate production have shown striking similarities in the basic pathway and thus these findings may provide strong clues regarding the mechanism in seaweeds [35, 36]. The molecular bases of alginate production begin with the fructose-6-phosphate and it is converted to guanosine di-phosphate-mannuronic acid (GDP-ManA) with a series of enzymatic transformations. Various enzymes including, mannose-6-phosphate isomerase (MPI), phosphomannomutase (PMM), mannose-1-phosphate guanylyltransferase (MPG), GDP-mannose/UDP glucose-6-dehydrogenase (GMD/UGD) are responsible for the synthesis of alginate precursor [35, 37]. GDP-ManA is then transferred across the cytoplasmic membrane and polymerized to the polymannuronate by the membrane-anchored proteins. After this stage, it may contain some residues unrelated to the alginate structure and it undergoes a modification step consisting of epimerization and degradation. The epimerization process is carried out by the mannuranoate C5-epimerases (MC5E) conducting the isomerization from mannuronic acid to guluronic acid. It should be underlined that the alginate synthesis route in bacteria differs from algae with the O-acetylation process that protects the produced alginate from degradation [38]. Finally, alginate polymer, composed of α-l-guluronic acid and the β-d-mannuronic acid, is formed, exported through the outer membrane, and released from the cell. Evidence for the biosynthesis of this polysaccharide within brown macroalgae come from a few studies with a limited number of species

Macroalgae species	Alginate yield (%)	References
Sargassum filipendula	17.2 ± 0.3	[22]
Sargassum vulgare	40	[23]
Sargassum wightii	33.18 ± 0.22	[21]
Sargassum angustifolium	3.5	[24]
Sargassum fluitans	9.36 ± 2.51	[25]
Sargassum muticum	13.57 ± 0.13	[26]
Sargassum natans	23 ± 1.6	[20]
Padina gymnospora	16 ± 0.7	
Padina antillarum	22 ± 1.1	
Laminaria digitata	29 ± 4.2	
Macrocystis pyrifera	26 ± 0.6	
Sargassum sp	31	[27]
Turbinaria sp	30	
Hormophysa sp	31	
Fucus spiralis	25 ± 0.21	[28]
Bifurcaria bifurcate	24 ± 0.12	
Ecklonia radiata	44 ± 0.15	[29]
Nizimuddinia zanardini	24 ± 0.8	[30]
Cystoseira barbata	9.9 ± 0.8	[31]
Padina pavonica	28.7	[32]
Ascophyllum nodosum	23.13	[33]
Durvillaea potatorum	55.2 ± 0.51	[34]
Seirococcus axillaris	41.3 ± 0.66	

Table 1.
The alginate content of various algae species.

such as *Ectocarpus siliculosus*, *Saccharina japonica* and *Laminaria digitate* [35, 37, 38]. Therefore, further research is needed to understand the metabolic route of alginate synthesis and to control the mechanism in different algae strains.

3. Extraction of alginate from algal material

Commercial alginate is mainly obtained from the biomass of brown macroalgae, and the conventional extraction process consists of multiple steps integrated to maximize product yield. Generally, the protocol begins with a pretreatment stage in which harvested and dried biomass is exposed to an acidic solution in order to break the cell wall, solubilize the relevant components, and reduce the viscosity of alginate to a desired level [26]. The second step is the alkali extraction, which is the most critical part of whole process because it greatly affects the yield and specific features of extracted alginate. At this stage, acidified biomass is treated with a strong alkali solution mostly sodium carbonate or sodium hydroxide in order to recover the alginic acid as soluble sodium alginate. The residue is removed with centrifugation or filtration and then the obtained extract is precipitated with the use of calcium chloride, hydrochloric acid, or sulfuric acid so as to precipitate alginates

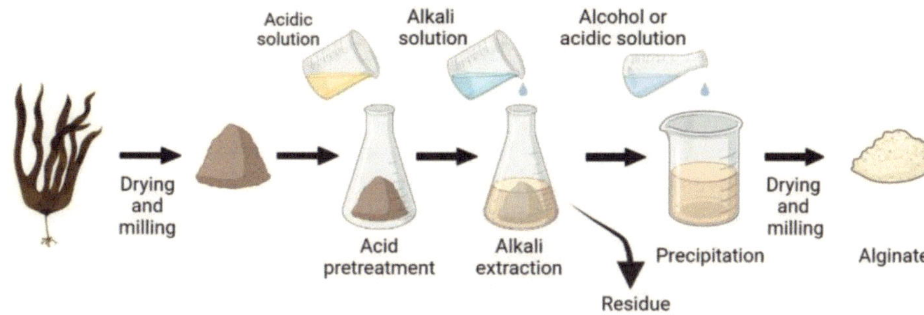

Figure 2.
Classical alginate extraction procedure from macroalgae.

in their acid or calcium salt form. Finally, the alginate product is dried, milled and ready for commercial use (**Figure 2**) [17, 39].

At the industrial level, the classical extraction method of alginate is widely used but it is highly complicated, time-consuming and requires high amount of solvents and chemicals. Therefore, novel approaches are suggested from several studies for the development of more suitable and effective extraction process (**Figure 3**). Sugiono et al. [40] performed an extrusion-assisted extraction procedure and optimized the key parameters (brown algae: solution ratio, feed rate and pH) for the alginate extraction from *Sargassum cristaefolium*. They reported that the extraction yield at optimum conditions reached the value of 34.96 ± 0.09%, and twin screw extruder was a promising method to extract alginate at the industrial scale. Youssouf et al. [41] proposed ultrasound-assisted extraction of alginate from *Sargassum muticum* to maximize extraction yield, minimize the use of chemicals, and shorten the process time. In another study, the deep eutectic solvent method combined with the subcritical water extraction technology were performed for the production of alginate from seaweed *Saccharina japonica*. The optimal conditions of different parameters were 150°C, 19.85 bar, 70% water content and 36.81 mL/g

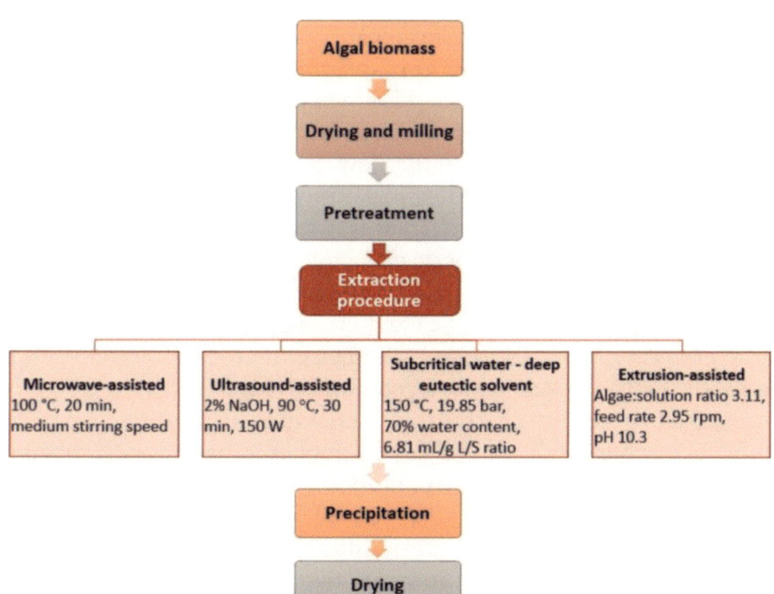

Figure 3.
Flow diagram of different extraction techniques from literature [40–43].

liquid/solid ratio giving an alginate yield of 28.1%. Also, the subcritical extraction method was defined as a clean, time-saving and effective process for the alginate extraction from seaweeds [42].

More recently, there has been a growing interest in the application of green technologies and biorefinery approach for the extraction of biological compounds. In this context, the development and optimization of biorefinery processes that integrate a sequential extraction steps in order to release multiple products of brown macroalgae is considered an effective, timesaving and green procedure. Several authors examined the extraction of a couple of components including alginate, fucoidan, laminarin, sugar, and so on with a biorefinery concept [33, 44, 45]. Yuan and Macquarrie [33] developed a step-by-step process to obtain a variety of products from *Ascophyllum nodosum* seaweed by the assistance of microwave technology. These products include fucoidan, alginates, sugars, and biochar (algae residue) and the obtained yields were 14.09, 18.24, 10.87, and 21.44% respectively. Kostas et al. [44] designed a bio-refinery procedure using *Laminaria digitata*, based on the extraction of the alginate and fucoidan, the subsequent production of bioethanol, and also the identification of bioactive compounds remaining in the residue. After the extraction of polysaccharides with the use of the conventional treatment method, the compositional structure of residue was analyzed and a high amount of glucose was determined, making this residue a potential feedstock for bioethanol production. This residue was exposed to acidic hydrothermal pretreatment and enzymatic saccharafication to release utilizable glucose and then it was fermented using *Saccharomyces cerevisiae* achieved an ethanol yield of 94.4%. Abraham et al. [45] developed and optimized a biorefinery process to extract polysaccharides of laminarin, fucoidan, and alginate from *Durvillaea potatorum*. The results established a novel biorefinery process for the extraction of multiple seaweed polysaccharides that could be used in specific industrial applications.

4. Immobilization of algae in alginate

Microalgae are one of the most remarkable species utilized in biotechnology for numerous purposes. They are crucial for biofuel production [46], bioremediation, and biotransformation [47], fuel cells applications [48], and also for wastewater treatment [49]. For this matter, the adaptation of efficient immobilization methods for microalgal applications is crucial to develop novel manufacturing strategies (**Table 2**). Most of these industries require low-cost and easy immobilization methods, of which alginate is one of the most profound encapsulation agents can serve this demand [63]. Additionally, due to their transparent nature, alginate matrices do not interfere with the photosynthetic efficiency of algae [64]. Various microalgae (*Chlamydomonas reinhardtii*, *Chlorella sp.*, *Botryococcus braunii*, *Tetraselmis sp.*, *Nannochloropsis sp.* and *Scenedesmus sp.*) and cyanobacteria (*Anabaena sp., Nostoc*, *Spirulina*, *Oscillatoria sp.*, etc.) species have been explored in immobilized matrix systems as beads, biofilms, and various geometries [61, 64, 65].

Biohydrogen as a green alternative fuel is known to be produced by microalgae species under anaerobic conditions [66]. Although microalgae are important for biohydrogen production, large-scale operations are hindered due to the oxygen sensitivity of hydrogenase enzymes [67]. Successful immobilization of *Chlamydomonas reinhardtti* and several other cyanobacteria species are promising to increase the biohydrogen production capacity of immobilized microalgae as densely packed biohydrogen micro factories [61, 62].

Microalgae in wastewater systems can also be immobilized with alginate for the continuous removal of nitrogen and phosphorous to decrease organic loads of

Target	Aim	Advantage	Disadvantage	Mode/Approach	Common microalgae species	References
Wastewater treatment/ Bioremediation	Removal of wastes and polluting chemicals	Reduced cost at downstream operations Enhanced cell survival Durable and long-term cultivation Continuous removal of nutrients, heavy metals, suspended solids and toxic organic compounds	Slower removal of phosphorus	Packed bed Airlift photobioreactors Biofilm photobioreactors Immobilized sheets Suspended alginate beads	*Scenedesmus dimorphus* *Chlorella vulgaris* *Spirulina platensis* *Chlorella sorokiniana* *Dunaliella salina*	[50–53]
Biotransformation	Decrease the toxic effect of compounds in aquatic systems Endocrine disrupting components Polycyclic aromatic hydrocarbons Phenolic compounds Dyes	Low capital cost High removal rates Small scale operations	Limited knowledge on microalgal biotransformation metabolism Requirement of extremophilic algae species Toxicity of the compounds to the algal cells	Suspended beads Immobilized alginate sheets Packed bed columns Airlift photobioreactors Biofilm photobioreactors Immobilized sheets	*Chlorella vulgaris* *Phormidium sp.* *Prototheca zopfii*	[54–57]
Biosensor	Environmental monitoring for aquatic and soil quality as toxicity bio-indication Suitable for agricultural and aquaculture purposes	Real time analysis Rapid pollutant removal	Disruption of alginate networks in water Dehydration and decomposition of the biosensor Lower detection quality	Disruption of PSII electron transfer by herbicides and other toxic molecules	*Chlorella vulgaris* *Chlamydomonas reinhardtii*	[58, 59]
Culture collection and handling	Increasing the success of long-term microalgae storage	Sustained metabolic activity at 4°C Time and cost efficient	Lack of standardized methods as it is in animal cell cultures	Immobilization in alginate beads or surfaces	Various important biotechnological species	[60]
Biohydrogen	Increasing the biohydrogen production efficiency and productivity	Prolonged hydrogen production compared to suspension cultures Enhanced cell viability in anaerobic cultures Decreased sensitivity to oxygen	Scale up Degradation of alginate in aqueous environment	Beads Sheets Tubular bioreactor Bubble column photobioreactor	*Chlamydomonas reinhardtii* *Anabaena sp.* *Synecocystis sp.* *Tetraspora sp.*	[61, 62]

Table 2.
Microalgae immobilization methods.

wastewater systems [53–55]. This approach is a clean and sustainable understanding for wastewater treatment, which inspired the utilization of microalgae for bioremediation purposes [68, 69] and removal of heavy metals [50] and other toxic molecules in the aquatic systems. There are also novel concepts to use immobilized microalgae networks as biosensors to check soil and water quality [58, 59].

Co-immobilization of different cell types can enhance the immobilized microalgae consortium. Microalgae growth-enhancing organisms can enhance the biomass accumulation in immobilized systems, which can increase the efficiency of immobilization for wastewater treatment, bioremediation, and biotransformation purposes [70, 71]. Symbiotic systems of algae-fungi in matrices can increase the efficacy of immobilization and decrease the toxic harms of heavy metals on algae [72].

Although alginate can provide a good environment for microalgae, there are several limitations concerning the stability of alginate gels. In aqueous systems, due to the diffusion of Ca^{+2} ions to aqueous environment, the alginate network can loosen, which subsequently damage the network. Thus, designer gels and/or blends with several other hydrogels can increase the durability and mechanical properties of alginate network [63, 73]. Another important aspect is although alginate does not affect cell proliferation, denser cultures may be needed, or due to dense culture diffusion limitation may increase cell death [62].

5. Algal alginate in food sector

Recently, food consumers have begun to consider nutrition contents of foods and desire more natural foods instead of the synthetic ones. As a result of that, foods which contain alginate as a natural substance have become more popular [74]. Most importantly, the United State Food and Drug Administration (U.S. FDA) has classified the alginate as "generally regarded as safe" and European Food Safety Authority (EFSA) has recognized to use it in specific doses [75, 76]. In the food sector, alginate has many uses as food production, packaging, thickening and, stabilizing agents, thanks to its unique properties like biodegradability, biocompatibility, renewability, and lack of toxicity [17, 74, 75, 77–79]. It has been noticed that alginate is easily tolerated in human body [80]. For this reason, it has been harmlessly inserted in a wide range of food products. Those can be listed as tinned, baked and, frozen foods, meat, poultry, salad, seafood, pet food, cheese, fruit, beverage, jelly, dessert, jam, ice cream, sorbet, and mayonnaise [17, 76, 81–84]. Additionally, it is considered functional food that has ability to reduce the risk of chronic diseases and make them more controllable [17, 85]. Thus, it enhances the quality of life due to its anticancer and probiotic features [17]. In addition, it can be applied to dairy liquid products, beer, and drinks which are consumed by diabetic patients [17, 80]. Also, adding alginate in the foods decreases the transit time in the colon and this situation helps human body to prevent from colon cancer [86, 87]. Moreover, as a result of having the ability to reduce the feeling of hunger, this polymer can be consumed to cure obesity [17, 86]. Moreover, it induces gastrointestinal disorders and the risk of coronary heart diseases [15, 87, 88]. Alginate is used in food products in the range of 0.5–1.5% [87]. For example, Na-alginate can be used without any unhealthful side effects at the highest dose of 15.5 mg Na-alginate/kg $(day)^{-1}$ [15]. Zn concentration should be carefully considered when Zn-alginate combination is used in food products. Because a high concentration of Zn^{2+} has negative effects on human body like nausea, diarrhea, and other diseases in the digestive system. So, its concentration must be in a suitable range. Zn-alginate can be added to purple corn to prevent the color in the drinks. Ca-alginate can be applied in yogurt, jams, and salads to control their smooth taste, in ice-cream to balance the crystal statement, and in noodles to increase the cohesion [78]. Propylene glycol-alginate can be included in salads and sauces [83]. Al^{3+} exhibit higher stable Al-alginate mixture than Ca^{2+} and Ba^{2+}, thanks to its three-dimensional binding model. But it is possibly toxic and is not safe for using in food products. Unfortunately, Al-alginate uses in food industry are limited as packaging material of conserve meals [78].

The food package is used for covering the product for protection, preservation, containment, and conservation purposes. After the food product is produced, physical/mechanical damages, physicochemical, and biological changes can occur. As a result, the quality and safety of the food may be decreased. In order to avoid this, synthetic compounds have begun to be used as a packaging material. Thereafter, it has been noticed that synthetic package materials are liable for a huge amount of waste that is detrimental to marine and wildlife. Therefore, researchers have been focused on finding new natural compounds that can be a promising candidate as a food packaging material [76]. After many experiments, they have been established that alginate has the ability to decrease lipid oxidation, microbial contaminations, nutrition lost, and wizening. Thus, this polymer improves the foods shelf life and keeps them fresh [76, 78, 89]. Nowadays, alginate is used for packaging in a wide range of food products like potato strips, pineapple, sweet cherry, peach, melon, pork, and beef balls, roast beef, chicken meat, chicken nugget, chicken ball, hams, salmon, bream, perch, mozzarella cheese, coffee, powdered milk, resoluble tea, fresh cut foods like apple, carrot, and mango [15, 76].

Nowadays, 3D food printing is an efficient technology to produce high valuable food products. While printing the food, encapsulation of significant compounds (antioxidants, vitamins, probiotics, etc.) with alginate increases the strength of foods against the negative effects of light, heat, and oxygen at preparation and storage stages. The most important problem in this regard is the tendency of food products to deteriorate geometrically. At this point, the alginate improves the water dispersion and thus provides more stable products with good mechanical and thermal behavior [74].

Alginate can be utilized as a good thickening agent, thanks to its adhesion and cohesion features. Pure alginate shows a high viscosity ten times more when compared to commercial thickeners. Also, it has the ability to enhance food properties like its texture, organoleptic situation, and consumer acceptance. For example, it can improve yogurt's shape, creamy texture, adhesion feature and restrain the viscosity at the sterilization step. Also, this polymer can be added to the jelly to decrease the difficulty involved in swallowing [78].

In food applications, there are many molecular surfactants that are used as a stabilizer; they have negative effects on human health and environment. As a result of this, researchers have been focused to find new solid particles that can be used instead of molecular surfactants. Solid particles are divided into two groups as inorganic and naturally derived. Unfortunately, inorganic particles have a limited area of usage [77]. Because of that, a rapid increase in the tendency to use surfactant derived from natural sources was observed [77, 90]. In this case, alginate can be added to the beer for stabilizing the foam as a stabilizing agent [78, 83]. Additionally, alginate can be mixed with oil droplets for the preparation of emulsion gels, which are used in mayonnaise and similar foods [78].

Alginate has the ability to combine with two different cations to form a gel. Alginate contained products have significant elasticity that is controllable by changing the ratios of ions and alginate concentrations [78, 91]. Besides having this unique property, algal alginate may include some impurities like heavy metals, polyphenols, proteins and endotoxins because it is a natural compound [17, 79]. In the food industry, low levels these impurities can be acceptable, but in the cosmetic industry, they have to be removed [79].

6. Algal alginate in the cosmetic industry

A cosmetic product can be defined as any natural or prepared material that in contact with teeth and mucous membranes of the mouth cavity and external parts of human body (epidermis, hair, nails, lips, and external genitals). These products can be in different forms as cream, lotion and spray. Nowadays, many people use cosmetic products and their ingredients, some for therapeutic purposes and others to enhance their beauty [92]. However, it should be noted that the purpose of using a cosmetic product cannot be to cure any parts of the human body. This kind of products are generally used after different dermatological issues like acne, eczema, and so on [93]. Recently, a new word called "cosmeceutical" has been used to indicate specific cosmetic products, which include active ingredients. These products are not considered drugs or cosmetics, but they show medical or drug-like benefits. The cosmetic/cosmeceutical consumers desire the products that are safe, effective, protective, elastic, and natural with good quality [91, 92].

Recently, cosmetic consumers have begun to pay attention not only to the effects of the product as a whole, but also to the content of the products [94]. With the increase in acquiring knowledge about the ingredients, awareness about the unhealthful side effects of synthetic cosmetic ingredients (irritation, allergic reactions, etc.) is created among the consumers more than before [91, 93, 94]. Additionally, Cosmetics Europe – the community for the cosmetics and personal care industry – has forbidden the use of synthetic solid plastic particles, which cannot be biodegradable by marine organisms for saving aquatic ecosystem in any types of cosmetic products [95]. This situation has contributed to conduct more research on finding new, natural, eco-friendly and biodegradable polymer sources to produce natural ingredients [91, 95]. At this point, algal alginate has drawn attention, thanks to its biological activities as an anticoagulant, antiviral, anticancer, antimicrobial, moisture retention, anti-irritating, antioxidant, anti-inflammatory, and antibacterial matter [17, 78, 90, 91]. As a result of having these aforementioned properties, alginate can be used as an abrasive agent, antioxidant, and thickening and stabilizing agents in the cosmetic industry [17, 90]. From this point of view, algal alginate is a promising candidate as a cosmetic ingredient.

The skin is the biggest organ of the human body and covers all the other organs [74, 91, 96]. It has three layers: epidermis, dermis, and hypodermis. The epidermis is composed of five stratums: basal, spinous, granular, lucid, and corneum. This layer contains melanocytes, langerhans, keratinocytes, granules, and dead keratinocytes [91, 93]. Melanocytes include melanin that determines the skin color and both of the melanocyte, and keratinocyte cells heal the skin damages. The stratum corneum acts like a water diffusion barrier. Thus, it protects the skin from dehydration and irritation and allows the human to live in air [91]. The health situation of the cells on the epidermis layer changes according to the weather conditions and nourishment schedules. Dead cells remain on the skin nearly for two weeks. After that, they through desquamation and recuperation stages and these stages take one month. Peeling products have the ability to remove dead cells and improve skin health without causing any negative effects on the skin. In this way, these products can help to make these processes faster [93]. Researchers have found that the optimum diameter of microparticles which is used in the peeling product formulation is 750 μm. Alginate microparticles are a good candidate as abrasive agents, thanks to their regular and spherical shape. The addition of starch to these microparticles increases the surface unevenness and inequality. This starch-alginate microparticles combination shows the effect on the skin as synthetic balls do. They have a unique potential for replacement with synthetic ones, as they are natural, biodegradable and environmentally friendly compounds [95].

Naturally the skin has the ability to synthesize antioxidant agents to protect itself from reactive oxygen species (ROS). Also, it has been known that the skin increases ROS production when exposed to UV radiation. Under these circumstances, oxidative stress causes the existence of wrinkles, dehydration, inflammation, melanoma, and skin cancer. For preventing skin aging and other aforementioned cutaneous disorders, the skin has to be supplied by antioxidants via cosmetic products [91, 93]. At this point, algal alginate is a promising candidate as a cosmetic ingredient with significant antioxidant activity [80, 91]. This activity related to its molecular weight, sulfate content and anionic groups [97]. Thus, it can be used as anti-aging, anti-wrinkle, and smoothing agents [93, 98]. Additionally, it has the ability to absorb several 100 times more water than its own weight to support the cell and regulate the water distribution in the skin, and thus protect cells from caving in [15, 91, 98]. Considering these properties, it has been inserted in a wide range of products such as hand lotions, ointments, fat-free creams, facial masks, and dental materials to improve nutrients diffusion and absorption [78, 80, 99].

Alginate can be used as thickening agent in shampoos, lotions, or other cosmetic products, which include huge amount of water for instability inhibition purpose [98]. Also, this polymer has ability to stabilize the viscosity to offer good liquidity in cosmetics [79]. This is the major reason of using it in cosmetic formulations [91]. Also, it helps to maintain the organoleptic features (taste, sight, smell, and touch) of cosmetics, thanks to its favorable activities [80, 91].

7. Algal alginate in pharmaceutical and biomedical applications

Although the biocompatibility of alginate has a debate, it is still one of the mostly studied polymeric biomaterials in pharmaceutical and biomedical applications for tissue engineering and regenerative medicine (TERM) purposes [100]. Alginate can be fabricated in various shapes and forms (**Figure 4**) for an extensively wide application (**Figure 5**). Alginate provides a biocompatible, cost-effective, low toxicity, and also easy gelation. Currently due to high viscosity and rheological properties with respect to increasing concentration, alginate is utilized as stabilizer and thickeners in pharmaceutical formulations. However, due to increased utilization of hydrogels in TERM, alginate-based formulations are extensively investigated as controlled drug-release platforms and tissue-engineering constructs [104–106].

Kinetic release of pharmaceutical compounds such as drug molecules, proteins, peptides, and nucleic acids is a novel advanced therapeutic approach [107]. Although alginate is a polar biopolymer, amphiphilic design of the alginate, or blending with other polymers can alter the hydrophilicity, thereby enabling the release of hypophobic/amphibic molecules [73, 103]. Alginate also creates a mild environment for proteins and other molecules, which can be affected by heat or alkali conditions resulting due to denaturation. Also, enzymes can be encapsulated with algae to have a controlled biocatalytic conversion. Alginate is usually ionically cross-linked with bivalent cations which is a low-cost and rapid method of gelation. However, when alginate is in an aqueous environment, bivalent cations are released into the environment, which makes a faster release of entrapped drug molecules based on their hydrophilicity, size, and interaction with alginate. In order to increase the control over the alginate, chemical modifications are done to chemically functionalize alginate for thermo-responsive, pH-responsive, or light-responsive matrices [108].

Wound healing is a complex phenomenon starting from inflammation, cell migration to the wound site, and eventually remodeling of the wound healing area [109]. Recently hydrogel-based wound dressings are gaining attention, and alginate

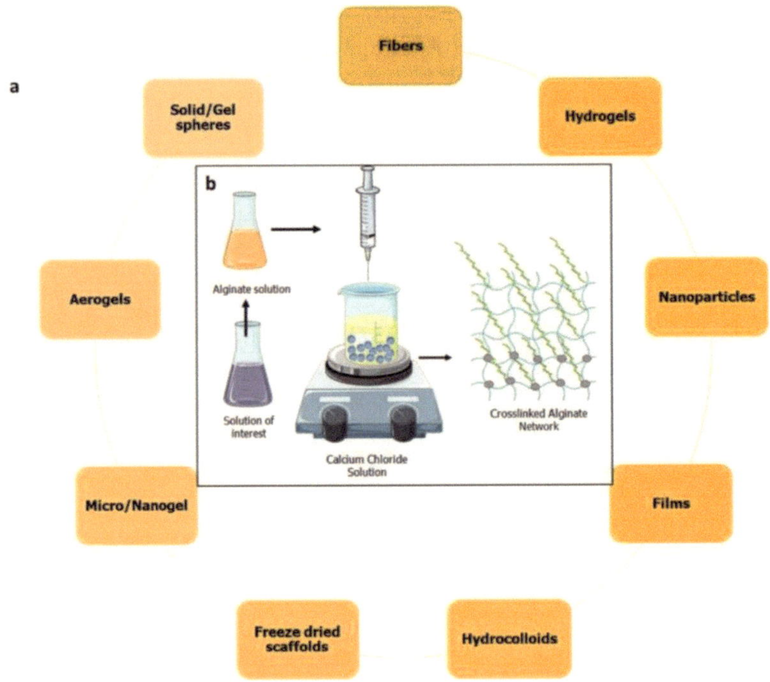

Figure 4.
Alginate in shape (a) fabrication forms of alginate for various application [61, 62, 100–102]; (b) classical immobilization method for alginate crosslinking.

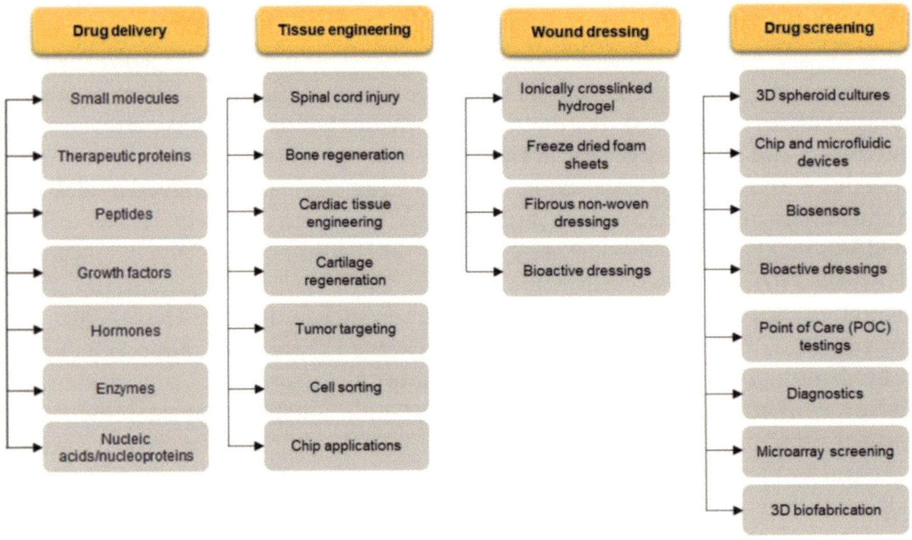

Figure 5.
Application areas of alginate in pharmaceutical and biomedical purposes [64, 100, 102, 103].

is one of the most studied and also commercially available wound dressing patches [103, 107]. Due to high water content and immobilization of bioactive molecules inside the patches to create antibacterial, anti-inflammatory and growth factors to promote cell growth and healing alginate are considered a gold standard in these types of applications.

3D cell culture is gaining interest because 2D cell culture does not correspond to the signals of cells in their nature. 3D environment creates a biomimetic environment to understand cell behavior, drug response, and 3D tissue culture [64, 102, 110]. Alginate creates a good environment resembling the extracellular matrix (ECM) structure where cells can proliferate and differentiate. Also, alginate can be covalently linked to cellular attachment sequences (mostly utilized RGD) to increase cell-cell interactions and cell-surface interactions [101, 111]. Encapsulation of growth factors in these 3D gels can increase the cell differentiation [104], neotissue formation, and blood vessel development [102, 112].

Alginate can also be a base hydrogel for 3D biofabrication purposes [111, 113]. Due to the availability of advanced imaging methods, these constructs can be customized as a personalized medicine tool [114]. However, due to the low mechanical properties of alginate, the bioprinting is usually done with blends with other hydrogels such as collagen [112], gelatin [111], chitosan [115] or self-assembling peptide hydrogels [104, 106].

Although alginate is a biocompatible and a plant-based biomaterial, the biodegradation of alginate can be troublesome. Alginate does not degrade in the body; however, due to the release of ions form the network, it decomposes into small pieces. Thus, chemical modification such as oxidation of alginate chains may help to achieve a proper biodegradation for clinical applications [116]. Moreover, low mechanical properties and stiffness may hinder the utilization of alginate, especially for hard tissue engineering. Chemical modification may elevate the material properties. However, it may add toxicity to the compound too. Nevertheless, as in vitro drug testing [117] and 3D cell culture platforms [111], even for topical applications [103], alginate is a safe natural biomaterial. It is also highly promising for tissue engineering applications, especially as injectable formulations [104].

8. Algal alginate in green nanotechnologies

Nanotechnology aims to have structures that have a size in a nanometer scale (less than 100 nm) to be produced and applied to provide purposeful design. Nanomaterials, which are a product of nanotechnology, have exceptional surface activity and other physical properties that occur due to their shapes at nanoscale sizes. In the last decade, nanotechnology has gained popularity and it has been used in different fields such as medicine, pharmaceuticals, cosmetics, food, and clothing industries. Production of synthetic nanomaterials is expensive and not an environmentally friendly process, even though they have many applications and benefits today. It is not safe to use them in medicinal applications due to their risks and side effects and the difficulty to form gels in situ. Hence, green routes to synthesize nanomaterials, which is called green nanotechnology has gained attention. The aim of green nanotechnology is to reduce the risks and to solve environmental problems related to nanotechnology [118, 119].

Natural polymers such as alginate, chitosan, agarose, collagen, cellulose, and so on have been used as nanoparticles (NPs) due to the concerns about synthetic ones [118]. Characteristics of these NPs such as small surface area to volume ratio, structural surfaces, agglomeration, and enhanced reactivity make them to be applied in various areas such as cancer therapy, drug targeting, nano-pharmacology, nanomedicine, and agrochemical delivery [120]. In recent times, the most widely used polymer is alginate, since it is considered safe especially for human applications. Alginate is considered to be safe owing to the fact that it has been studied extensively, even though other biomaterials can be good alternatives in the future, However, alginate has properties that offer advantages to the system and make it a perfect fit for biotechnology and drug delivery systems via cell microencapsulation [118]. Temperature and pH changes, signaling molecules, and enzymes stimulate a drastic chemical and physical change in alginate, which results in making them a potential candidate for drug delivery vehicles [121]. Biocompatible and nontoxic polyionic complex NPs are formed through ionic gelation of alginate and chitosan. These polyionic complexes are used in drug delivery and wound healing purposes because they are non-toxic and biocompatible as well as have effective protection of biomolecules [122]. Natural nano carrier systems can be easily integrated with antiviral, antifungal, antituberculosis drugs, and so on. For antituberculosis drugs, lipid-based formulations and polymer-based formulations are used. Lipid-based formulations have drawbacks with successful targeting, since it is dependent on the parenteral/inhalable route, whereas alginate is already FDA approved for human use and it is successful with the oral treatment of reflux esophagitis as well as being a popular pharmaceutical excipient. Hence, alginate-based carriers have gained popularity in drug targeting. The recent studies prove that if alginate NPs are used, the outcome could be further improved in the sense of encapsulation of drug, pharmacokinetics, bioavailability, and therapeutic efficacy [123]. Alginate NPs can also be used as a carrier for adjuvants and vaccine immunogenicity is increased, since alginate nanocarriers can prolong the release. Agglomeration has not occurred in major organs through the use of alginate NPs. Mucoadhesive properties enhance the permeability of alginate NPs and therefore it is being used in nasal and oral administrations; degradation is reduced in acidic environment [124].

NPs of alginate can be used in agriculture as a nanopesticide, nanoinsecticide, nanoherbicide, nanofertilizer, growth stimulants, pesticide carriers, antimicrobial agents, and nanoformulations [125]. Targeting and systemic delivery of herbicides can be provided by using nanocapsules with alginate/chitosan NPs [126]. Chitosan and alginate as carriers of herbicide and insecticide do not only improve the release of the herbicide but also improves its interaction with the soil [126, 127]. Chitosan/alginate NPs can also be used as nano carriers for pesticides, herbicides, and fungicides. Slow release of the molecule can be provided and NPs can protect them from UV radiation and it offers a better antifungal activity [120].

9. Algal alginate as a coagulant in wastewater treatment

Coagulation is a process used in water treatment, in which aids are used to change the surface structure of suspended materials to form aggregates and to remove them by destabilization. In this process, inorganic metals and polymers are generally used as coagulation aids [128–130]. The large amount of chemicals, significant pH changes, and the high amount of sludge produced are among the significant disadvantages of the coagulation process with metal salts [128, 129]. In addition, some negative effects of synthetic polymer on human health have increased the tendency to use natural polymeric materials as a coagulation agents. Natural polymeric materials such as polysaccharides are low cost, easy to obtain, have low molecular weight and high shear stability. For these reasons, they have been suggested to be more advantageous materials. They also have advantages such as being safe for human health, biodegradable, and having a wider effective flocculation dose range for various colloidal suspensions [129, 131]. High volume wastewater is one of the most important problems for many industrial sectors. Especially, textile industries produce large volumes of wastewater with varying physicochemical properties. This diversity in physicochemical properties is due to the enormous continuous effort to identify suitable technologies for the treatment of textile industry wastewater and the many components involved in this process [130, 132]. The different types of wastewater treatment performed for industrial wastewater include coagulation/flocculation, oxidation, membrane separation,

ion exchange, photochemical, adsorption, biological treatment method, and so on [130, 133]. Among the various methods, one of the effective methods for removing substances from wastewater is coagulation using algal alginate [130].

Alginate naturally derived from algae offers significant potential for wastewater treatment as a coagulant. Calcium and sodium ions can be used as coagulation aids in processes where alginate is used as a coagulant. Especially when calcium ions interact with metal cations in the alginate structure, the gel structure forms and tends to precipitate the pollution factors in the wastewater. Thanks to having the ability of formation insoluble molecules, it becomes an important option as a coagulant in wastewater treatment [130, 134–136].

Laboratory-scale studies on the use of the obtained algal alginate in wastewater treatment processes have been carried out. In these studies, the process continues with measuring the coagulation efficiency depending on the determined parameters after the extraction stage. In the study conducted by Vijayaraghayan and Shanthakumar [130], *Sargassum sp.* was used as an alginate source and the efficiency of removing Congo red dye from the aqueous solution was studied depending on the pH, alginate dose, calcium dose, and initial dye concentration of the extracted alginate. As a result of the study, it has been shown that the performance of alginate as a coagulant is highly dependent on the calcium dose used as the gelling agent and the initial dye concentration in the solution [130]. A process for reactive magenta dye removal in textile wastewater was carried out depending on the alginate dose, calcium dose, and pH by the same authors. In this study, a color removal of 92.7% was achieved and it was confirmed that the alginate extracted from *Sargassum sp.* could be used as a coagulant for dye removal in textile wastewater [137]. In another study conducted by Natesh et al., 3 different algae such as *Sargassum* sp., *Turbinaria* sp., and *Kaapaphycous alvarezii* were used and the results supported the study obtained by Vijayaraghayan and Shanthakumar [129]. In the study by Devrimci et al., the coagulation efficiency of algal alginate was investigated in terms of drinking water treatment. The study was carried out depending on parameters such as calcium and alginate doses, and the initial turbidity of the samples. Experiments on synthetically prepared turbid water samples have shown that calcium alginate can act as a potential coagulant. The coagulation efficiency was highly dependent on the initial turbidity and calcium concentration. At high initial turbidity, the coagulant worked well, and the targeted final turbidity was achieved at most doses of calcium and alginate. It is stated that the performance is weaker at low turbidity. The authors noted that the use of higher viscosity alginate and prolonged rapid mixing may improve the performance for low turbidity waters [135].

10. Conclusions

Algae are considered a major source of alginate. Since their alginate content and properties are varying, first, the amount to be used should be decided. According to this decision, algae species and the time for growing and harvesting of them must be taken into attention. After that, depending on the area of use, the extraction method should be determined in order to obtain the highest yield of alginate from algal biomass. Now, it is ready to be utilized in different types of sectors. For example, immobilized microalgae networks are open to novel applications. Environmental monitoring and algae-based biosensors comprise one of the promising topics for future developments. Rather than classical bead or thin-film fabrication methods, novel biofabrication techniques can be adapted for algae immobilization, which can help to design customized geometries. Also, as a result of the ability to combine with two different cations to form gel, alginate-contained products show significant elasticity. Unfortunately, algal alginate may contain some impurities like heavy metals, polyphenols, proteins, and endotoxins. In the food industry, low levels of them can be acceptable. But before they are used in cosmetic and pharmaceutic industries, they have to be removed using some purification methods. Alginate NPs have properties such as being biocompatible, nontoxic, and biodegradable. They are safe and preparation of the alginate NPs is easy and so this makes them a potential carrier for drug delivery systems. They can be applied to various drug-delivery systems. Alginate NPs are FDA approved as a food additive and has great mucoadhesive properties, which can make them a potential candidate for drug delivery through the oral route. In agriculture, chitosan/alginate NPs are used mostly for targeted and systemic delivery of agrochemicals, and it has a great potential for prolonged availability and low load of the molecules. Agricultural technology and increase of the fertilizers and pesticides unfortunately made a negative contribution to environment. However, NPs, especially "green NPs," made agriculture more sustainable by using lower doses and slower release of the molecules. Increased awareness of the environmental problems comes with an unavoidable sustainability in all fields, and green NPs are good for environment because their application in agriculture is safe, and also their productions are considered sustainable. However, there are certain limitations to the industrial application of alginate. The most important of these limitations is the exponentially increasing cost with growing scale. For example, coagulants currently used for wastewater treatment are relatively cheaper than algal alginate. However, traditional coagulation processes may require extra costly processes such as pH adjustment or alkalinity addition. Today, in studies about algal alginate, it is possible to increase the efficiency of the system and reduce the cost of coagulation with alginate by using their better and more suitable quality versions.

Author details

Cagla Yarkent, Bahar Aslanbay Guler, Ceren Gurlek, Yaprak Sahin, Ayse Kose, Suphi S. Oncel and Esra Imamoglu*
Faculty of Engineering, Department of Bioengineering, Ege University, Izmir, Turkey

*Address all correspondence to: esraimamoglu@yahoo.com

References

[1] Shanmugam H, Sathasivam R, Rathinam R, Arunkumar K, Carvalho IS. Algal biotechnology: An update from industrial and medical point of view. Omics Technologies and Bio-Engineering. 2018;**2**:31-52. DOI: 10.1016/B978-0-12-815870-8.00003-6

[2] Rao NRH, Tamburic B, Doan YTT, Nguyen BD, Henderson RK. Algal biotechnology in Australia and Vietnam: Opportunities and challenges. Algal Research. 2021;**56**(May):102335. DOI: 10.1016/j.algal.2021.102335

[3] Padina gymnospora (Kützing) Sonder 1871: Algaebase [Internet]. 2021. Available from: https://www.algaebase.org/search/species/detail/?tc=accept&species_id=4577

[4] Ecklonia radiata (C. Agardh) J. Agardh 1848: Algaebase [Internet]. 2021. Available from: https://www.algaebase.org/search/species/detail/?species_id=3463

[5] Sargassum [Internet]. 2021. Available from: https://stringfixer.com/tr/Sargassum

[6] Galatchi M, Nenciu M, Tania Z. Viata in Marea Neagra. 2014

[7] Macrocystis pyrifera. Wikipedia [Internet]. 2021. Available from: https://en.wikipedia.org/wiki/Macrocystis_pyrifera

[8] Laminaria digitata [Internet]. 2021. Available from: https://www.seaweed.ie/descriptions/Laminaria_digitata.php

[9] Nature Picture Library Cabbage coral field {Turbinaria sp} Papua New Guinea – Juergen Freund [Internet]. 2021. Available from: https://www.naturepl.com/stock-photo-cabbage-coral-field-turbinaria-sp-papua-new-guinea-nature-image01141630.html

[10] Nature Picture Library Knotted wrack (Ascophyllum nodosum) Wester Ross, Scotland – Niall Benvie [Internet]. 2021. Available from: https://www.naturepl.com/stock-photo/knotted-wrack-(ascophyllum-nodosum)-wester-ross-scotland/search/detail-0_01658116.html

[11] Goh CH, Heng PWS, Chan LW. Alginates as a useful natural polymer for microencapsulation and therapeutic applications. Carbohydrate Polymers. 2012;**88**(1):1-12. DOI: 10.1016/j.carbpol.2011.11.012

[12] Venkatesan J, Nithya R, Sudha PN, Kim SK. Role of alginate in bone tissue engineering. Advances in Food and Nutrition Research. 2014;**73**:45-57. DOI: 10.1016/B978-0-12-800268-1.00004-4

[13] Kraan S. Algal polysaccharides, novel applications and outlook. Carbohydrates – Comprehensive Studies on Glycobiology and Glycotechnology. 2012;(May):489-532. DOI: 10.5772/51572

[14] Titlyanov EA, Titlyanova TV, Li X, Huang H. Marine plants of coral reefs. Coral Reef Marine Plants of Hainan Island. 2017:5-39. DOI: 10.1016/b978-0-12-811963-1.00002-0

[15] Puscaselu RG, Lobiuc A, Dimian M, Covasa M. Alginate: From food industry to biomedical applications and management of metabolic disorders. Polymers (Basel). 2020;**12**(10):1-30

[16] Doyle WT. Brown Algae. Nonseed Plants: Form and Function. 1970:127-137. DOI: 10.1016/B978-0-12-741550-5.50023-4

[17] Łabowska MB, Michalak I, Detyna J. Methods of extraction, physicochemical properties of alginates and their applications in biomedical field – A review. Open Chemistry. 2019;**17**(1):738-762

[18] Li Y, Fu X, Duan D, Xu J, Gao X. Comparison study of bioactive substances and nutritional components of brown algae Sargassum fusiforme strains with different vesicle shapes. Journal of Applied Phycology. 2018; **30**(6):3271-3283

[19] Öztaşkent C, Ak İ. Effect of LED light sources on the growth and chemical composition of brown seaweed Treptacantha barbata. Aquaculture International. 2021;**29**(1):193-205

[20] Rhein-Knudsen N, Ale MT, Ajalloueian F, Meyer AS. Characterization of alginates from Ghanaian brown seaweeds: Sargassum spp. and Padina spp. Food Hydrocolloids. 2017;**71**:236-244. DOI: 10.1016/j.foodhyd.2017.05.016

[21] Kumar S, Sahoo D. A comprehensive analysis of alginate content and biochemical composition of leftover pulp from brown seaweed Sargassum wightii. Algal Research. 2017;**23**:233-239. DOI: 10.1016/j.algal.2017.02.003

[22] Bertagnolli C, Espindola APDM, Kleinübing SJ, Tasic L, Silva MGCD. Sargassum filipendula alginate from Brazil: Seasonal influence and characteristics. Carbohydrate Polymers. 2014;**111**:619-623. DOI: 10.1016/j.carbpol.2014.05.024

[23] Sari-Chmayssem N, Taha S, Mawlawi H, Guégan JP, Jeftić J, Benvegnu T. Extracted and depolymerized alginates from brown algae Sargassum vulgare of Lebanese origin: Chemical, rheological, and antioxidant properties. Journal of Applied Phycology. 2016;**28**(3):1915-1929

[24] Borazjani NJ, Tabarsa M, You SG, Rezaei M. Effects of extraction methods on molecular characteristics, antioxidant properties and immunomodulation of alginates from Sargassum angustifolium. International Journal of Biological Macromolecules. 2017;**101**:703-711. DOI: 10.1016/j.ijbiomac.2017.03.128

[25] Davis D, Simister R, Campbell S, Marston M, Bose S, McQueen-Mason SJ, et al. Biomass composition of the golden tide pelagic seaweeds Sargassum fluitans and S. natans (morphotypes I and VIII) to inform valorisation pathways. Science of the Total Environment. 2021;**762**:143134. DOI: 10.1016/j.scitotenv.2020.143134

[26] Mazumder A, Holdt SL, De Francisci D, Alvarado-Morales M, Mishra HN, Angelidaki I. Extraction of alginate from Sargassum muticum: Process optimization and study of its functional activities. Journal of Applied Phycology. 2016;**28**(6):3625-3634. DOI: 10.1007/s10811-016-0872-x

[27] Rahelivao MP, Andriamanantoanina H, Heyraud A, Rinaudo M. Structure and properties of three alginates from madagascar seacoast algae. Food Hydrocolloids. 2013;**32**(1):143-146. DOI: 10.1016/j.foodhyd.2012.12.005

[28] Bouissil S, El Alaoui-Talibi Z, Pierre G, Michaud P, El Modafar C, Delattre C. Use of alginate extracted from Moroccan brown algae to stimulate natural defense in date palm roots. Molecules. 2020;**25**(3):1-15. DOI: 10.3390/molecules25030720

[29] Lorbeer AJ, Lahnstein J, Bulone V, Nguyen T, Zhang W. Multiple-response optimization of the acidic treatment of the brown alga Ecklonia radiata for the sequential extraction of fucoidan and alginate. Bioresource Technology. 2015;**197**:302-309. DOI: 10.1016/j.biortech.2015.08.103

[30] Khajouei RA, Keramat J, Hamdami N, Ursu AV, Delattre C, Laroche C, et al. Extraction and characterization of an alginate from the Iranian brown seaweed Nizimuddinia zanardini. International Journal of Biological Macromolecules. 2018;**118**:1073-1081. DOI: 10.1016/j.ijbiomac.2018.06.154

[31] Sellimi S, Younes I, Ayed HB, Maalej H, Montero V, Rinaudo M, et al. Structural, physicochemical and antioxidant properties of sodium alginate isolated from a Tunisian brown seaweed. International Journal of Biological Macromolecules. 2015;**72**:1358-1367. DOI: 10.1016/j.ijbiomac.2014.10.016

[32] Faidi A, Farhat F, Boina DA, Touati M, Le-Nouen D, Stumbé JF. Physico-chemical characterization of alginates isolated from a Tunisian Padina pavonica algae as a sustainable biomaterial. Polymer International. 2020;**69**(11):1130-1139

[33] Yuan Y, Macquarrie DJ. Microwave assisted step-by-step process for the production of fucoidan, alginate sodium, sugars and biochar from Ascophyllum nodosum through a biorefinery concept. Bioresource Technology. 2015;**198**:819-827. DOI: 10.1016/j.biortech.2015.09.090

[34] Lorbeer AJ, Charoensiddhi S, Lahnstein J, Lars C, Franco CMM, Bulone V, et al. Sequential extraction and characterization of fucoidans and alginates from Ecklonia radiata, Macrocystis pyrifera, Durvillaea potatorum, and Seirococcus axillaris. Journal of Applied Phycology. 2017;**29**(3):1515-1526

[35] Chi S, Liu T, Wang X, Wang R, Wang S, Wang G, et al. Functional genomics analysis reveals the biosynthesis pathways of important cellular components (alginate and fucoidan) of Saccharina. Current Genetics. 2018;**64**(1):259-273

[36] Moradali MF, Ghods S, Rehm BHA. Alginate Biosynthesis and Biotechnological. Production. 2018:1-25

[37] Nyvall P, Corre E, Boisset C, Barbeyron T, Rousvoal S, Scornet D, et al. Characterization of mannuronan C-5-Epimerase genes from the brown alga laminaria digitata. Plant Physiology. 2003;**133**(2):726-735

[38] Michel G, Tonon T, Scornet D, Cock JM, Kloareg B. Central and storage carbon metabolism of the brown alga Ectocarpus siliculosus: Insights into the origin and evolution of storage carbohydrates in Eukaryotes. The New Phytologist. 2010;**188**(1):67-81

[39] Mohammed A, Rivers A, Stuckey DC, Ward K. Alginate extraction from Sargassum seaweed in the Caribbean region: Optimization using response surface methodology. Carbohydrate Polymers. 2020; **245**(June):116419. DOI: 10.1016/j.carbpol.2020.116419

[40] Sugiono S, Masruri M, Estiasih T, Widjanarko SB. Optimization of extrusion-assisted extraction parameters and characterization of alginate from brown algae (Sargassum cristaefolium). Journal of Food Science and Technology. 2019;**56**(8):3687-3696. DOI: 10.1007/s13197-019-03829-z

[41] Youssouf L, Lallemand L, Giraud P, Soulé F, Bhaw-Luximon A, Meilhac O, et al. Ultrasound-assisted extraction and structural characterization by NMR of alginates and carrageenans from seaweeds. Carbohydrate Polymers. 2017;**166**:55-63. DOI: 10.1016/j.carbpol.2017.01.041

[42] Saravana PS, Cho YN, Woo HC, Chun BS. Green and efficient extraction of polysaccharides from brown seaweed by adding deep eutectic solvent in subcritical water hydrolysis. Journal of Cleaner Production. 2018;**198**:1474-1484. DOI: 10.1016/j.jclepro.2018.07.151

[43] Silva M, Gomes F, Oliveira F, Morais S, Delerue-Matos C. Microwave-assisted alginate extraction from Portuguese Saccorhiza polyschides – Influence of acid pretreatment, world academy of science, engineering and

technology, open science index 97. International Journal of Biotechnology and Bioengineering. 2015;**9**(1):30-33

[44] Kostas ET, White DA, Cook DJ. Development of a bio-refinery process for the production of speciality chemical, biofuel and bioactive compounds from Laminaria digitata. Algal Research. 2017;**28**(November):211-219. DOI: 10.1016/j.algal.2017.10.022

[45] Abraham RE, Su P, Puri M, Raston CL, Zhang W. Optimisation of biorefinery production of alginate, fucoidan and laminarin from brown seaweed Durvillaea potatorum. Algal Research. 2019;**38**(August):101389. DOI: 10.1016/j.algal.2018.101389

[46] Kose A, Oncel SS. Algae as a promising resource for biofuel industry: facts and challenges. International Journal of Energy Research. 2016:941-951

[47] Deviram G, Mathimani T, Anto S, Ahamed TS, Ananth DA, Pugazhendhi A. Applications of microalgal and cyanobacterial biomass on a way to safe, cleaner and a sustainable environment. Journal of Cleaner Production. 2020;**253**:119770. DOI: 10.1016/j.jclepro.2019.119770

[48] Wu JY, Lay CH, Chia SR, Chew KW, Show PL, Hsieh PH, et al. Economic potential of bioremediation using immobilized microalgae-based microbial fuel cells. Clean Technologies and Environmental Policy. 2021;(0123456789). DOI: 10.1007/s10098-021-02131-x

[49] Ma J, Wang Z, Zhang J, Waite TD, Wu Z. Cost-effective Chlorella biomass production from dilute wastewater using a novel photosynthetic microbial fuel cell (PMFC). Water Research. 2017;**108**:356-364. DOI: 10.1016/j.watres.2016.11.016

[50] Ahmad A, Bhat AH, Buang A. Biosorption of transition metals by freely suspended and Ca-alginate immobilised with Chlorella vulgaris: Kinetic and equilibrium modeling. Journal of Cleaner Production. 2018;**171**:1361-1375

[51] Bayramoğlu G, Tuzun I, Celik G, Yilmaz M, Arica MY. Biosorption of mercury(II), cadmium(II) and lead(II) ions from aqueous system by microalgae Chlamydomonas reinhardtii immobilized in alginate beads. International Journal of Mineral Processing. 2006;**81**(1):35-43

[52] García De Llasera MP, Santiago ML, Flores EJL, Bernal Toris DN, Covarrubias Herrera MR. Mini-bioreactors with immobilized microalgae for the removal of benzo(a)anthracene and benzo(a) pyrene from water. Ecological Engineering. 2018;**121**:89-98

[53] Jiménez-Pérez MV, Sánchez-Castillo P, Romera O, Fernández-Moreno D, Pérez-Martínez C. Growth and nutrient removal in free and immobilized planktonic green algae isolated from pig manure. Enzyme and Microbial Technology. 2004;**34**(5): 392-398

[54] Ertuğrul S, Bakir M, Dönmez G. Treatment of dye-rich wastewater by an immobilized thermophilic cyanobacterial strain: Phormidium sp. Ecological Engineering. 2008;**32**(3):244-248

[55] Gao QT, Wong YS, Tam NFY. Removal and biodegradation of nonylphenol by immobilized Chlorella vulgaris. Bioresource Technology. 2011;**102**(22):10230-10238. DOI: 10.1016/j.biortech.2011.08.070

[56] Sol A, Matamoros V. Removal of Endocrine Disrupting Compounds from Wastewater by Microalgae Co-immobilized in Alginate Beads. 2016;

[57] Wu H, Fu Y, Guo C, Li Y, Jiang N, Yin C. Electricity generation and removal performance of a microbial fuel cell using

sulfonated poly (ether ether ketone) as proton exchange membrane to treat phenol/acetone wastewater. Bioresource Technology. 2018;**260**(April):130-134

[58] Ferro Y, Perullini M, Jobbagy M, Bilmes SA, Durrieu C. Development of a biosensor for environmental monitoring based on microalgae immobilized in silica hydrogels. Sensors (Switzerland). 2012;**12**(12):16879-16891

[59] Perullini M, Ferro Y, Durrieu C, Jobbágy M, Bilmes SA. Sol-gel silica platforms for microalgae-based optical biosensors. Journal of Biotechnology. 2014;**179**(1):65-70. DOI: 10.1016/j.jbiotec.2014.02.007

[60] Chen Y-C. Immobilization of Twelve Benthic Diatom Species for Long-Term Storage and as Feed for Post-Larval Abalone Haliotis Diversicolor. Aquaculture. 2007;**263**(1-4):97-106. DOI: 10.1016/j.aquaculture.2006.12.008

[61] Kosourov S, Murukesan G, Seibert M, Allahverdiyeva Y. Evaluation of light energy to H2 energy conversion efficiency in thin films of cyanobacteria and green alga under photoautotrophic conditions. Algal Research. 2017;**28**(April):253-263. DOI: 10.1016/j.algal.2017.09.027

[62] Canbay E, Kose A, Oncel SS. Photobiological hydrogen production via immobilization: Understanding the nature of the immobilization and investigation on various conventional photobioreactors. Biotechnology. 2018;**8**(5):0. DOI: 10.1007/s13205-018-1266-3

[63] Moreira SM, Moreira-Santos M, Guilhermino L, Ribeiro R. Immobilization of the marine microalga Phaeodactylum tricornutum in alginate for in situ experiments: Bead stability and suitability. Enzyme and Microbial Technology. 2006;**38**(1-2):135-141

[64] Lode A, Krujatz F, Brüggemeier S, Quade M, Schütz K, Knaack S, et al. Green bioprinting: Fabrication of photosynthetic algae-laden hydrogel scaffolds for biotechnological and medical applications. Engineering in Life Sciences. 2015;**15**(2):177-183. DOI: 10.1002/elsc.201400205

[65] Solé A, Matamoros V. Removal of endocrine disrupting compounds from wastewater by microalgae co-immobilized in alginate beads. Chemosphere. 2016;**164**:516-523

[66] Oncel SS. Biohydrogen from microalgae, uniting energy, life, and green future. Handbook of Marine Microalgae: Biotechnology Advances. 2015:159-196. DOI: 10.1016/B978-0-12-800776-1.00011-X

[67] Oncel SS, Kose A, Faraloni C, Imamoglu E, Elibol M, Torzillo G, et al. Biohydrogen production using mutant strains of Chlamydomonas reinhardtii: The effects of light intensity and illumination patterns. Biochemical Engineering Journal. 2014;**92**:47-52. DOI: 10.1016/j.bej.2014.06.019

[68] Bozarth A, Maier UG, Zauner S. Diatoms in biotechnology: Modern tools and applications. Applied Microbiology and Biotechnology. 2009;**82**(2):195-201

[69] Shimoda K, Hamada H. Bioremediation of Bisphenol a and Benzophenone by Glycosylation with Immobilized Marine Microalga Pavlova sp. Environ Health Insights. 2009;**3**:89-94

[70] De-Bashan LE, Hernandez JP, Morey T, Bashan Y. Microalgae growth-promoting bacteria as "helpers" for microalgae: A novel approach for removing ammonium and phosphorus from municipal wastewater. Water Research. 2004;**38**(2):466-474

[71] Gonzalez LE, Bashan Y. Increased growth of the microalga Chlorella vulgaris when coimmobilized and cocultured in alginate beads with the plant-growth-promoting bacterium

Azospirillum brasilense. Applied and Environmental Microbiology. 2000; **66**(4):1527-1531

[72] Araji MT, Shahid I. Symbiosis optimization of building envelopes and micro-algae photobioreactors. Journal of Building Engineering. 2018; **18**(February):58-65. DOI: 10.1016/j.jobe.2018.02.008

[73] Çelik E, Bayram C, Akçapinar R, Türk M, Denkbaş EB. The effect of calcium chloride concentration on alginate/Fmoc-diphenylalanine hydrogel networks. Materials Science and Engineering: C. 2016;**66**:221-229

[74] Mallakpour S, Azadi E, Hussain CM. State-of-the-art of 3D printing technology of alginate-based hydrogels – An emerging technique for industrial applications. Advances in Colloid and Interface Science. 2021;**293**:102436. DOI: 10.1016/j.cis.2021.102436

[75] Cathell MD, Szewczyk JC, Schauer CL. Organic modification of the polysaccharide alginate. ChemInform. 2010;**41**(24):61-67. DOI: 10.1002/chin.201024253

[76] Kontominas MG. Use of alginates as food packaging materials. Food. 2020;**9**(10):1-5. DOI: 10.3390/foods9101440

[77] Daradmare S, Choi KH, Kim J, Park BJ. Preparation of eco-friendly alginate-based Pickering stabilizers using a dual ultrasonic nebulizer spray method. Journal of Industrial and Engineering Chemistry. 2020;**84**:96-105. DOI: 10.1016/j.jiec.2019.12.025

[78] Hu C, Lu W, Mata A, Nishinari K, Fang Y. Ions-induced gelation of alginate: Mechanisms and applications. International Journal of Biological Macromolecules. 2021;**177**:578-588. DOI: 10.1016/j.ijbiomac.2021.02.086

[79] Sanchez-Ballester NM, Bataille B, Soulairol I. Sodium alginate and alginic acid as pharmaceutical excipients for tablet formulation: Structure-function relationship. Carbohydrate Polymers. 2021;**270**(July):118399. DOI: 10.1016/j.carbpol.2021.118399

[80] Pereira L, Cotas J. Introductory chapter: Alginates – A general overview. Alginates – Recent Uses This Natural Polymer. 2020;(February)

[81] Bord Iascaigh Mhara. Scoping a Seaweed Biorefinery Concept For Ireland Report for Bord Iascaigh Mhara. 2020

[82] DG Mare, Easme. Blue Bioeconomy Forum: [Internet]. 2019. Available from: https://op.europa.eu/en/publication-detail/-/publication/c8b2f69f-4314-11ea-b81b-01aa75ed71a1/language-en

[83] McLachlan J. Macroalgae (seaweeds). Industrial Resources and Their Utilization. 2021;**89**(1):227-241

[84] Mushollaeni W. The physicochemical characteristics of sodium alginate from Indonesian brown seaweeds. African Journal of Food Science. 2011;**5**(6):349-352

[85] Qin Y. Research, Development and Commercial Applications of Bioactive Seaweed Substances in Health Products. Annals of Public Health & Epidemiology. 2020;**1**(4):1-9. DOI: 10.33552/APHE.2020.01.000516

[86] Lähteenmäki-Uutela A, Rahikainen M, Camarena-Gómez MT, Piiparinen J, Spilling K, Yang B. European Union legislation on macroalgae products. Aquaculture International. 2021;**29**(2):487-509

[87] Seitz A, Reed S, Weakley C. Alginates: Handling/Processing. Technical Evaluation Report Complied by OMRI for the USDA National Organic Program. 2015:1-23

[88] Solah VA, Kerr DA, Adikara CD, Meng X, Binns CW, Zhu K, et al.

Differences in satiety effects of alginate- and whey protein-based foods. Appetite. 2010;**54**(3):485-491. DOI: 10.1016/j.appet.2010.01.019

[89] Juraj J, Technology F, Kuhača F. Macroalgae in the food industry. Opportunities and Challenges. 2017:14-19

[90] Covis R, Vives T, Gaillard C, Benoit M, Benvegnu T. Interactions and hybrid complex formation of anionic algal polysaccharides with a cationic glycine betaine-derived surfactant. Carbohydrate Polymers. 2015;**121**:436-448. DOI: 10.1016/j.carbpol.2015.01.001

[91] Wang HMD, Chen CC, Huynh P, Chang JS. Exploring the potential of using algae in cosmetics. Bioresource Technology. 2015;**184**:355-362

[92] Yarkent Ç, Gürlek C, Oncel SS. Potential of microalgal compounds in trending natural cosmetics: A review. Sustainable Chemistry and Pharmacy. 2020;**17**(February):1-11. DOI: 10.1016/j.scp.2020.100304

[93] Lourenço-Lopes C, Fraga-Corral M, Jimenez-Lopez C, Pereira AG, Garcia-Oliveira P, Carpena M, et al. Metabolites from macroalgae and its applications in the cosmetic industry: A circular economy approach. Resources. 2020;**9**(9):1-30. DOI: 10.3390/RESOURCES9090101

[94] Morais T, Cotas J, Pacheco D, Pereira L. Seaweeds compounds: An ecosustainable source of cosmetic ingredients? Cosmetics. 2021;**8**(1):1-28

[95] Kozlowska J, Prus W, Stachowiak N. Microparticles based on natural and synthetic polymers for cosmetic applications. International Journal of Biological Macromolecules. 2019; **2019**(129):952-956

[96] Zhang M, Qiao X, Han W, Jiang T, Liu F, Zhao X. Alginate-chitosan oligosaccharide-ZnO composite hydrogel for accelerating wound healing. Carbohydrate Polymers. 2021;**266**(April):118100. DOI: 10.1016/j.carbpol.2021.118100

[97] Ruocco N, Costantini S, Guariniello S, Costantini M. Polysaccharides from the marine environment with pharmacological, cosmeceutical and nutraceutical potential. Molecules. 2016;**21**(5):1-16

[98] Kumari R, Joshi S, Upasani VN, Student BS. Applications of algae in cosmetics: An overview. International Journal for Innovative Research in Science & Technology. 2018;**7**(2): 1269-1278

[99] Hernández-Carmona G, Freile-Pelegrín Y, Hernández-Garibay E. Conventional and alternative technologies for the extraction of algal polysaccharides. Functional Ingredients from Algae for Foods and Nutraceuticals. 2013:475-516

[100] Drury JL, Mooney DJ. Hydrogels for tissue engineering: Scaffold design variables and applications. Biomaterials. 2003;**24**:4337-4351. DOI: 10.1016/S0142-9612(03)00340-5

[101] Kirdponpattara S, Khamkeaw A, Sanchavanakit N, Pavasant P, Phisalaphong M. Structural modification and characterization of bacterial cellulose-alginate composite scaffolds for tissue engineering. Carbohydrate Polymers. 2015;**132**:146-155. DOI: 10.1016/j.carbpol.2015.06.059

[102] Maiullari F, Costantini M, Milan M, Pace V, Chirivì M, Maiullari S, et al. A multi-cellular 3D bioprinting approach for vascularized heart tissue engineering based on HUVECs and iPSC-derived cardiomyocytes. Scientific Reports. 2018;**8**(1):1-15

[103] Lee KY, Mooney DJ. Alginate: Properties and biomedical applications. Progress in Polymer Science. 2011;**37**(1):

106-126. DOI: 10.1016/j.progpolymsci.2011.06.003

[104] Ghosh M, Halperin-Sternfeld M, Grinberg I, Adler-Abramovich L. Injectable alginate-peptide composite Hydrogel as a scaffold for bone tissue regeneration. Nanomaterials. 2019;**9**(4):2-3

[105] Gong X, Branford-White C, Tao L, Li S, Quan J, Nie H, et al. Preparation and characterization of a novel sodium alginate incorporated self-assembled Fmoc-FF composite hydrogel. Materials Science and Engineering: C. 2016;**58**:478-486

[106] Xie Y, Zhao J, Huang R, Qi W, Wang Y, Su R, et al. Calcium-ion-triggered co-assembly of peptide and polysaccharide into a hybrid hydrogel for drug delivery. Nanoscale Research Letters. 2016;**11**(1)

[107] Dexter AF, Fletcher NL, Creasey RG, Filardo F, Boehm MW, Jack KS. Fabrication and Characterization of Hydrogels Formed from Designer Coiled-Coil Fibril-Forming Peptides. Royal Society of Chemistry Advances. 2017;7:27269-27271. DOI: 10.1039/c7ra02811c

[108] Soledad Lencina MM, Iatridi Z, Villar MA, Tsitsilianis C. Thermoresponsive hydrogels from alginate-based graft copolymers. European Polymer Journal. 2014;**61**:33-44

[109] Yun WJ, Bang SH, Min KH, Kim SW, Lee MW, Chang SE. Epidermal growth factor and epidermal growth factor signaling attenuate laser-induced melanogenesis. Dermatologic Surgery. 2013;**39**(12):1903-1911

[110] Mondal A, Gebeyehu A, Miranda M, Bahadur D, Patel N, Ramakrishnan S, et al. Characterization and printability of Sodium alginate - Gelatin hydrogel for bioprinting NSCLC co-culture. Scientific Reports. 2019;**9**(1):1-12

[111] Ouyang L, Yao R, Zhao Y, Sun W. Effect of bioink properties on printability and cell viability for 3D bioplotting of embryonic stem cells. Biofabrication. 2016;**8**(3):1-12. DOI: 10.1088/1758-5090/8/3/035020

[112] Calvo Catoira M, Fusaro L, Di Francesco D, Ramella M, Boccafoschi F. Overview of Natural Hydrogels for Regenerative Medicine Applications. Journal of Materials Science: Materials in Medicine. 2019;**30**(10):1-10. DOI: 10.1007/s10856-019-6318-7

[113] Wang Y, Huang X, Shen Y, Hang R, Zhang X, Wang Y, et al. Direct Writing Alginate Bioink Inside Pre-Polymers of Hydrogels to Create Patterned Vascular Networks. Journal of Materials Science. 2019;**54**(10);7883-7892. DOI: 10.1007/s10853-019-03447-2

[114] Sonnaert M, Papantoniou I, Bloemen V, Kerckhofs G, Luyten FP, Schrooten J. Human periosteal-derived cell expansion in a perfusion bioreactor system: proliferation, differentiation and extracellular matrix formation. Journal of Tissue Engineering and Regenerative Medicine. 2017;**11**(2):519-530

[115] Han J, Zhou Z, Yin R, Yang D, Nie J. Alginate-chitosan/hydroxyapatite polyelectrolyte complex porous scaffolds: Preparation and characterization. International Journal of Biological Macromolecules. 2010;**46**(2):199-205

[116] Boontheekul T, Kong HJ, Mooney DJ. Controlling alginate gel degradation utilizing partial oxidation and bimodal molecular weight distribution. Biomaterials. 2005;**26**(15):2455-2465

[117] Miranda JP, Rodrigues A, Tostões RM, Leite S, Zimmerman H,

Carrondo MJT, et al. Extending hepatocyte functionality for drug-testing applications using high-viscosity alginate-encapsulated three-dimensional cultures in bioreactors. Tissue Engineering, Part C: Methods. 2010;**16**(6):1223-1232

[118] Ibrahim M, Salman M, Kamal S, Rehman S, Razzaq A, Akash SH. Algae-based biologically active compounds. Algae Based Polymers, Blends, and Composites. 2017:155-271

[119] Nasrollahzadeh M, Sajjadi M, Sajadi SM, Issaabadi Z. Green nanotechnology. Interface Science and Technology. 2019;**28**:145-198. DOI: 10.1016/B978-0-12-813586-0.00005-5

[120] Agrahari S, Dubey A. Nanoparticles in plant growth and development. Biog Nano-Particles Their Use Agro-Ecosystems. 2020:9-37

[121] Kanwar R, Rathee J, Salunke DB, Mehta SK. Green nanotechnology-driven drug delivery assemblies. ACS Omega. 2019;**4**(5):8804-8815

[122] Douglas KL, Piccirillo CA, Tabrizian M. Effects of alginate inclusion on the vector properties of chitosan-based nanoparticles. Journal of Controlled Release. 2006;**115**(3):354-361

[123] Ahmad Z, Pandey R, Sharma S, Khuller GK. Alginate nanoparticles as antituberculosis drug carriers: Formulation development, pharmacokinetics and therapeutic potential. The Indian Journal of Chest Diseases & Allied Sciences. 2006;**48**(3):171-176

[124] Saraei F, Mohamadpour Dounighi N, Zolfagharian H, Moradi Bidhendi S, Khaki P, Inanlou F. Design and evaluate alginate nanoparticles as a protein delivery system. Archives of Razi Institute. 2013;**68**(2):139-146

[125] Gupta A, Bano A, Rai S, Pathak N, Sharma S. New insights into application of nanoparticles for plant growth promotion: Present and future prospects. Biog Nano-Particles Their Use Agro-Ecosystems. 2020:259-279

[126] Kumar N, Balamurugan A, Shafreen MM, Rahim A, Vats S, Vishwakarma K. Nanomaterials: Emerging trends and future prospects for economical agricultural system. Biog Nano-Particles Their Use Agro-Ecosystems. 2020:281-305

[127] Silva M d S, Cocenza DS, Grillo R, Melo NFS d, Tonello PS, Oliveira LC d, et al. Paraquat-loaded alginate/chitosan nanoparticles: Preparation, characterization and soil sorption studies. Journal of Hazardous Materials. 2011;**190**(1-3):366-374. DOI: 10.1016/j.jhazmat.2011.03.057

[128] Nozaic DJ, Freese SD, Thompson P. Longterm experience in the use of polymeric coagulants at Umgeni water. Water Science and Technology: Water Supply. 2001;**1**(1):43-50

[129] Prasanna Natesh P, Sricharan V, Iyankumar R, Vishnu PD. A novel method of algal based water treatment by natural coagulant "Alginates.". International Journal of Science and Research. 2017;**6**(9):769-775

[130] Vijayaraghavan G, Shanthakumar S. Performance study on algal alginate as natural coagulant for the removal of Congo red dye. Desalination and Water Treatment. 2016;**57**(14):6384-6392

[131] Kawamura S. Effectiveness of natural polyelectrolytes in water treatment. Journal American Water Works Association. 1991;**83**(10):88-91

[132] Eremektar G, Selcuk H, Meric S. Investigation of the relation between COD fractions and the toxicity in a textile finishing industry wastewater: Effect of preozonation. Desalination. 2007;**211**(1-3):314-320

[133] Chuah TG, Jumasiah A, Azni I, Katayon S, Thomas Choong SY. Rice husk as a potentially low-cost biosorbent for heavy metal and dye removal: An overview. Desalination. 2005;**175**(3): 305-316

[134] Çaylak B, Vardar SF. Comparison of different production processes for bioethanol. Turkish Journal of Chemistry. 1998;**22**(4):351-359

[135] Devrimci HA, Yuksel AM, Sanin FD. Algal alginate: A potential coagulant for drinking water treatment. Desalination. 2012;**299**:16-21. DOI: 10.1016/j.desal.2012.05.004

[136] Shelar PS et al. Medicinal value of seaweeds and its applications–A Review. Continental Journal of Pharmacology and Toxicology Research. 2012;**5**(2):1-22

[137] Seenuvasan M, Suganthi JRG, Sarojini G, Malar GCG, Priya ME, Kumar MA. Effective removal of reactive magenta dye in textile effluent by coagulation using algal alginate. Desalination and water treatment. Desalin. Water Treatment. 2018;**121**:22-27

Section 2

Therapy-Based Studies

Chapter 6

Alginate-Based Composite and Its Biomedical Applications

Yaling Deng, Ningning Yang, Oseweuba Valentine Okoro, Amin Shavandi and Lei Nie

Abstract

Alginate has received much attention due to its biocompatibility. However, the properties of pure alginate are limited, such as weak mechanical strength, which limits its application. Alginate-based composite effectively overcomes the defect of pure alginate. The molecular weight and microstructure can be designed. More importantly, the essential properties for clinical application are improved, including mechanical properties, biocompatibility, gelation ability, chondrogenic differentiation and cell proliferation. This chapter will describe development of alginate-based composite in biomedical application. In the fields of wound dressing, drug delivery, and tissue engineering, the impact of structural changes on performance has been stated. To provide readers with understanding of this chapter, the structure and characterization of alginate will be included.

Keywords: Alginate, composite, wound dressing, drug delivery, tissue engineering

1. Introduction

Alginate is a natural heteropolysaccharide extracted from brown seaweed. Due to the abundantly available in nature and less expensive, *L. hyperborea*, *L. digitata*, Laminaria japonica, *A. nodosum*, and *M. pyrifera* are the main choice of alginate [1]. Alginate, without pungent smell, usually has a white or yellowish-brown character. It is water-soluble but insoluble in organic solvents, such as alcohol, chloroform. Carboxylic acid groups in the saccharide residue endow the special anionic nature. Calcium alginate, for instance, is insoluble, while sodium alginate has water-solution [2]. However, the process of water dissolving is slow and usually takes several hours, forming stable and viscous solutions [3]. Due to the ability to form a hydrogel, alginate could be utilized as a gelling agent, which expands its application in the field of biomedical applications.

Alginate and its derivatives show outstanding properties, such as biocompatibility, biodegradation, gel-forming ability, being suitable for sterilization and storage. Owing to its unique characteristics, alginate has been widely used in diverse fields, including wound dressings, drug delivery, tissue regeneration. In this chapter, the characteristics and wide applications of alginate will be introduced.

2. Structure and characterization

2.1 Structure

Alginate with molecular weight between 32,000 to 400,000 g/mol is mainly comprised of a sequence of liner polymers of β-(1–4)-D-mannuronic (M-blocks), α-L-guluronic acid (G-blocks), and inserted MG sequences (MG-blocks), with varying proportions and linear arrangements [4]. That organized in homogenous patterns with repeated G residues, repeated M residues, and heterogenous patterns with alternating G and M residues [5]. Alginate derived from different sources displays different M/G ratios and contents in M and G [6], leading to the change of molecule weight and physicochemical properties. These parameters are related to the characteristics and applications of alginate. Generally, alginate with high M units shows good biocompatibility and more immunogenic [7]. Alginate with high M units has soft and elastic properties, G-rich alginate exhibits hard and brittle characteristics [8, 9]. The rigidity of the chains increases in a sequence, MG < MM < GG, due to the electrostatic repulsion between charged groups. G-rich alginate gels have better mechanical stability (**Figure 1**) [5].

The electrostatic interactions between the carboxylate groups of G units and divalent cations, such as Ca^{2+}, Fe^{2+}, Mg^{2+}, form an "egg-box" structure, which crosslinks to obtain the hydrogels. There are several hydroxy and carboxyl groups in the molecular structure of alginate. As the active site, more groups and side-chain molecules were introduced into the main chain to decorate the structure, which endow more features and expand its applications [10].

2.2 Solubility and viscosity

Sodium alginate exhibits slowly water-soluble and forms viscous and stable solution. At low solvent pH, more heterogeneous MG-blocks contribute to solute

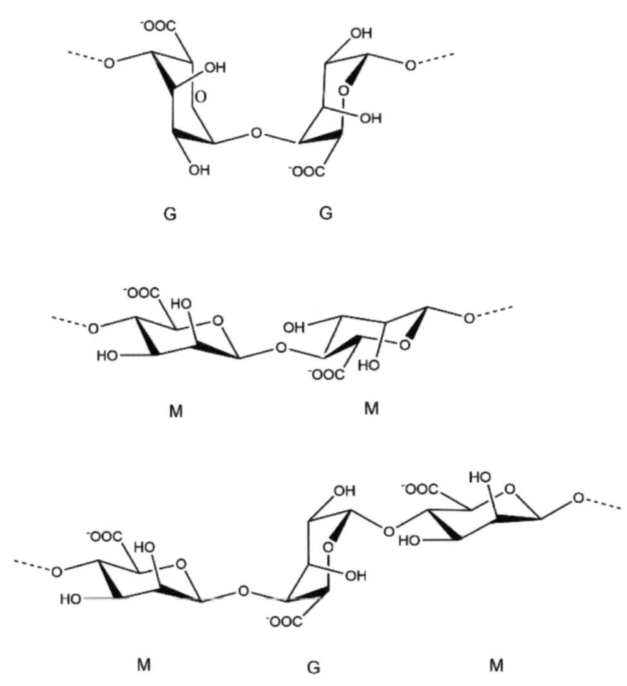

Figure 1.
Chemical structure of alginate [5].

than M-rich and G-rich alginate [11]. With the decrease of pH, the viscosity of alginate solution increase. When the range of pH is 3–3.5, the viscosity has the maximum. In the Mark-Houwink relationship ($[\eta] = KM_v^a$), $[\eta]$ is intrinsic viscosity (mL/g) and M_v is the viscosity-average molecular weight (g/mol)[12]. For sodium alginate in 0.1 M NaCl solution at 25°C, K and a, the parameter of the Mark-Houwink relationship ($[\eta] = KW_v^a$), are 2×10^{-3}, $a = 0.97$ respectively.

The primary structure is associated with different amounts and sequential distribution of M and G, which affects the molecular weight and properties. For example, the viscous behavior is remarkably relevant to molecular weight during the preparation. Alginate with high molecular weight polymer could form an obviously viscous solution [13] and result in a higher elastic modulus in the gels [14]. Meanwhile, the alginate with a long chain shows a higher solution viscosity. For example, G-rich alginate displays more excellent water solubility than M-rich alginate [15].

2.3 Biocompatibility

Alginate has excellent biocompatibility, that has been widely assessed. However, some gaps of biocompatibility between alginate and alginate complex still exist. For purified alginate gels, the relative amounts and distribution of M and G have an effect on biocompatibility [16]. For example, M. Otterlei et al. reported that alginates with low G were approximately 10 times more potent inducing cytokine production compared with high G alginates [7]. S.K. Tam et al. found that gel beads prepared with alginate contenting intermediate guluronate (IntG, 44% G) exhibited better biocompatible than high guluronate content (HiG, 71% G). There are no inflammatory reactions around alginate implants [17]. For the alginate complex, the impurities from alginate-based materials, such as heavy metals, proteins, and polyphenolic compounds, have the potential to cause an immunogenic response. A multi-step extraction procedure reduces the concentration of impurities and will not cause foreign body reactions. Meanwhile, the biocompatibility properties of alginate are attributed to hydrophilicity, chain migration, and water-absorbing [18]. Swelling properties contribute to enhance biocompatibility, that limiting the adsorption of proteins and cells of immune response [19].

2.4 Degradation

Degradability of biomaterials, as a critical property, contributes to providing a biomimetic microenvironment for use in cell delivery, survival, and expansion [20]. Hydrolytic is the main reason for degradation, which will be initiated spontaneously after contacting water-based fluids [21]. The degradation rate is related to the molecular weight of alginate. Generally, with the increase of molecular weight, the amounts of reactive sites for hydrolysis degradation decrease, that reduce the degradation rate [22]. Degradation, accompanied by adjusting the structure and molecular weight distribution, has an effect on the mechanical properties. In the physiological environment, alginate will undergo rapid degradation [23]. For alginate hydrogel, the degradation process will be accomplished by releasing the divalent ions into the surrounding environment. However, some residues of alginate still exist and will not be removed [24]. Many physical and chemical methods have been used to control alginate degradation, including ultrasonic, ultraviolent, gamma irradiation, and partial oxidation.

3. Applications

The various physicochemical properties of alginate are favorable to extensive usages, such as the ability to form a gel and the ability to maintain a moist environment, which has been confirmed by the literature. Due to the biocompatibility and nontoxic, alginate has been applied in biomedical applications, such as wound dressing, drug delivery, and tissue engineering.

3.1 Wound dressing

Skin, as the largest human organ, plays a vital role in protecting the human body from outside germs [25]. It is also a vulnerable organ. Once the skin is damaged, all kinds of microbes and pathogens will gather around the wound site to affect the wound closure cause bacterial infection. For severe wounds with large, deep, or bleeding, handling and management are essential [26]. Wound healing is a complex process, involving the treatment of infectious, germs. Therefore, wound dressing has been attracted extensive attention and is widely used to accelerate wounding recovery and to control infection. Traditional dressing, consisting of sterile pad and gauze, has the features of preventing the invasion of germ into the wound site and keeping dry [27]. Some disadvantages of these dressings are displayed during the application, such as poor vapor transmission, susceptiblity to bacterial infection, and easy adhesion, which seriously interfere with wound healing. With the rapid development of technology, some biomaterials such as polysaccharides, proteoglycans, and proteins have been investigated for wound healing owing to their ability to accelerate healing and control infection [28, 29]. These polymers have been utilized to develop modern dressing due to its high water-content, biodegradable, and biocompatibility. Therefore, modern wound dressing effectively mimics the features of natural tissue and offers a moist environment, better water vapor permeation, autolytic debridement, which cause to promote re-epithelialization and accelerate wound healing [30].

Alginate, as one of these biomaterials, has been widely concentrated and investigated because of its biocompatibility and water-retaining capacity [31]. It has the ability to carry pro-inflammatory signals and initiate or accelerate the healing of chronic wounds [32]. Currently, there are several forms of alginate dressing on the market, including hydrogel, films, membrane, and sponges. Ionic cross-linking is a representative fabrication method of alginate dressings. Multivalent ions are chosen to fabricate gel, such as calcium [33], magnesium [34], iron [35]. Free-dried porous sheets and fibrous dressing that followed are formed. Alginate dressings are characterized by absorbing wound fluid, physiologically moist environment, maintain the optimal pH, and reduced bacterial infections, that improve the physicochemical stability [36, 37].

Alginate wound dressings with multifunctional properties have been investigated for nearly 40 years. Kaltostat® alginate dressing, as a famous wound dressing, has been designed and applied to absorb exudate and avoid infections, especially calcium sodium alginate dressing. Nevertheless, only 70% wound closure is exhibited [38]. Considering that single alginate dressing could not completely prevent infections and moderate wound healing, the development of effective alginate dressing, based on biopolymers and nanoparticles, draw much attention and expanded research.

Alginate composites combining natural polymer and synthetic polymer are utilized to prepare dressing hydrogel via physical or chemical methods. Chitosan, as a typical natural polymer, has several advantages, such as biocompatibility, biodegradability, ionic character, and hemostatic properties. It could be incorporated with alginate to create bioactive interpolymer (polyelectrolyte) complexes and promising wound dressing. Wound dressing formed with this method exhibited superior water uptake of 4343.4% over 24 h and 78% biodegradation than commercial Pharma-Algi wound dressing [39]. Mudlovu et al. fabricated an alginate-chitosan bioplatform through a three-step method; partial-crosslinking of polymers, lyophilization, and pulverization (**Figure 2**) [39]. These wound dressings showed a higher degree and rate of fluid uptake (3306.61%, 4343.4%) than Pharma-Algi® (2168.21%; 1612.56%). For chitosan with high (MW ~ 800 000) and low (MW ~ 3000) molecular weight, it was mixture as a coating to develop calcium alginate dressing. The results showed that chitosan promotes angiogenesis by increasing VEGF and exhibits better wound healing than pure calcium alginate dressing [40]. In addition to the natural polymer, poly (vinyl alcohol) (PVA), as a type of common synthetic polymer, has the inherent characteristics of non-toxic, non-carcinogenic, biocompatibility. PVA hydrogel formed with cross-linking method has desirable properties such as high swelling rate, keeping the moist environment. Therefore, PVA has been attracted attention and explored for wound dressing. PVA-alginate hydrogel enhanced swelling properties and protein adsorption. When alginate has a high ratio in the PVA/Calcium alginate wound dressing, high a water vapor transmission rate, 2725.8 $g/m^2/2h$, was obtained, which contributes to holding a moist environment [41].

In order to endow the additional effects and accelerate wound healing, the antimicrobial agent has been incorporated into the wound dressing. Silver nanoparticles (AgNPs), as a promising wide-spread local antibacterial agent, have demonstrated huge potential and received more attention in the area of microbial resistance. Wound dressing carried with AgNPs is considered to control and treat acute and chronic wounds, accompanied by rapid healing via accelerating reepithelialization [42]. Alginate dressings incorporated with silver have highly absorbent features of alginate and antimicrobial efficacy of silver. Silver-loaded hydrogel has a uniform pore structure, had excellent water absorption and water retention, maintaining a moist wound environment for wounds [43]. The antibacterial performance is also

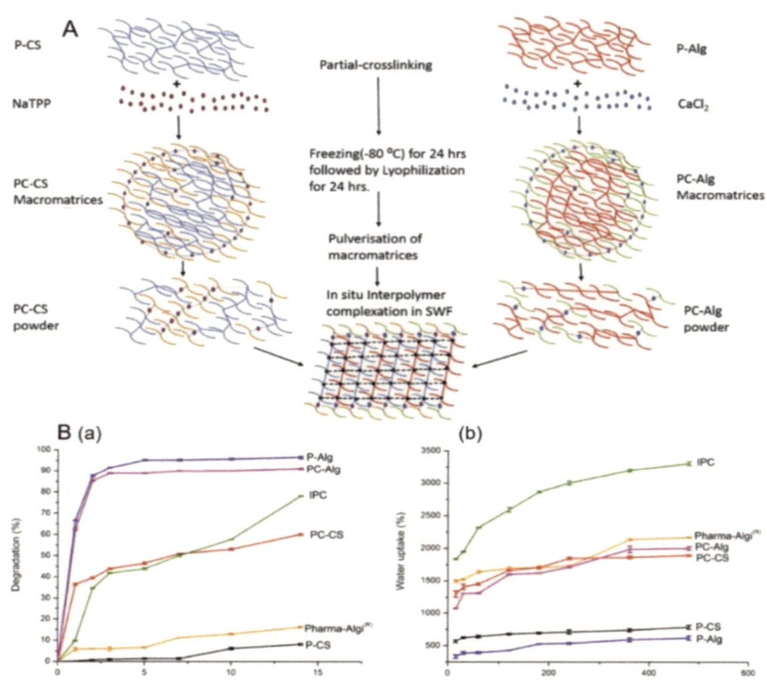

Figure 2.
(A) Schematic representation of the three-step method of partial-crosslinking and interpolymer complexation of the polymers. (B) (a) biodegradation and (b) water uptake behavior of the pristine polymers, bioplatforms and pharma-Algi® in PBS (pH 7.4) solution at 37°C at 50 rpm [39].

improved. Alginate/carboxymethyl cellulose silver dressing was found to have a sustained antimicrobial effect against microorganisms, for up to 21 days [44]. The sustain and stable release of sliver minimize clinical treatment time and relieve the patient's pain. Compared with pure alginate, alginate dressings incorporating ionic silver and nanocrystalline silver were confirmed to enhanced antimicrobial effect, improved the binding for elastase, matrix metalloproteases-2, and proinflammatory cytokines, and boosted the antioxidant capacity [45].

3.2 Drug delivery

Drug delivery systems could transport and release drug molecules to a target position under the microenvironment, especially for poorly water-soluble drug molecules. Many polymers have been explored and investigated for drug delivery. Due to the outstanding biocompatibility and gelation properties, alginate has been received wide attraction and utilized in drug delivery applications. Alginate could be prepared excipient. Three forms of participation are exhibited, such as solid dosage, semisolid dosage, liquid dosage. Solid dosage form contains tablets and capsules, while, gels and buccal patches are the most common semisolid dosage. Liquid dosage usually includes emulsions and suspensions. Alginate could be decorated to adjust the properties for novel drug release. Due to the facility to bond with drug molecules and rapid gelation in a mild environment, alginate could be utilized to modulate the drug release properties. D.J. Mooney et al. found that the pore size of alginate gel is about 5–6 nm [46], which contributes to the diffusion and release of drug molecules.

Alginate could be utilized to design microcapsules for sustained release of the drug, which has attracted extraordinary attention by the reason of outstanding advantages, including high drug loading rate, satisfactory biodegradable, non-immunogenic, and non-toxicity. Stimuli-responsive alginate-based microcapsules had shown a potential for targeted delivery and release drugs. Drug delivery system using magnetic nanoparticles was reported. Magnetic nanoparticles have the ability to control movement and aggregation at the target sites during the process of the external magnetic field. Therefore, the system could obtain a high local concentration at the target sites. It is demonstrated that magnetic reduction-responsive alginate-based microcapsules (MRAMCs) systems exhibited good magnetic targeted ability owing to the superparamagnetism of OA-Fe_3O_4 nanoparticles [47]. In addition, alginate microcapsules encapsulated glucocorticoid is a novel method for the treatment of osteoarthritis. *E. Lengert* et al. found that hollow silver alginate microcapsules could effectively deliver water-insoluble glucocorticoid betamethasone [48]. This system enhanced loading efficiency and sustained release, reducing the injury rate of intraarticular glucocorticoids delivery.

Alginate, as a well-known drug carrier, is utilized to prepare wound dressing. Antimicrobial agents or drugs are encapsulated into the hydrogel. Wound dressing provides a moist environment, and more importantly, drugs release to the wound and facilitate the healing. For example, alginate-based dressing encapsulated with vancomycin has a 44% drug release rate after 24 h, and the antimicrobial activity against various bacteria was confirmed [49]. Recently, double-membrane hydrogel formed with alginate and cellulose nanocrystals was developed. The results showed good release behaviors for complexing antibiotic drugs. The rapid drug release was completed by outer neat alginate hydrogel, meanwhile, the prolonged-release behavior corresponded to the inner hydrogel [50]. Another dual-drug delivery system, poly (D, L-lactic) (PDLLA) microspheres embedded in calcium alginate hydrogel beads, was developed by D.G. Zhong et al. [51]. The microspheres encapsulated glycyrrhetinic acid showed a sustained release, and hydrogel loaded with BSA exhibited a rapid release.

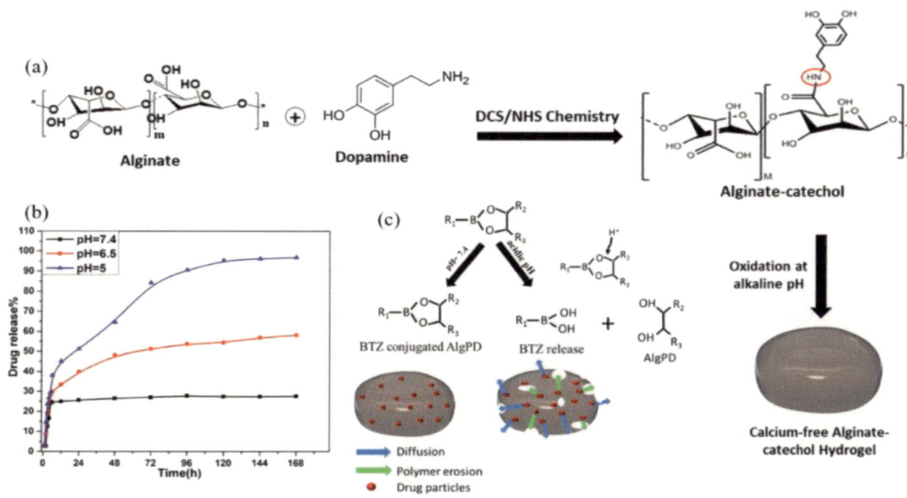

Figure 3.
The synthesis of AlgPD-BTZ hydrogel and drug release properties. (a) Schematic of the synthesis process of AlgPD-BTZ hydrogel by the conjugation of dopamine to the alginate backbone, and subsequent oxidation of catechol groups for cross-linking; (b) pH sensitive cumulative drug release from AlgPD-BTZ hydrogel; (c) scheme shows pH sensitive boronic ester bond of BTZ with the polymeric catechol group and the drug release mechanism [52].

pH-sensitive alginate carrier was investigated to enhance the efficacy, especially for localized drug delivery systems. The pH-sensitive reversible bond can control the drug delivery. C. H. Park and C. S. Kim prepared AlgPD-BTZ hydrogel using alginate-conjugated polydopamine as a building block polymer (**Figure 3**) [52]. The catechol group binds to the boronic acid group of BTZ drug, that covalent bond is a pH-sensitive reversible bond. Therefore, the system showed that BTZ was selectively released in cancer cells with a pH-dependent method. Sodium alginate could be used as a pH-sensitive bilayer coating on iron oxide nanoparticles by combining hydroxyapatite to deliver drug molecules. The higher encapsulation efficiency was detected, 93.03 ± 0.23% for curcumin and 98.78 ± 0.05% for 6-gingerol [53]. The electrostatic interactions of molecules could act as an adjustable gate to hold and release the drug molecules depending on the pH. The pH triggered drug-releasing mechanism play a virtual role in the releasing of tumor drug owing to the leaky vasculature [54].

In addition, alginate hydrogels are an excellent candidate and have been studied for protein drug delivery. Protein incorporated into hydrogels is protected, which reduces the denaturation and avoids degradation. Alginate hydrogels containing adjusting factors of neovascularization, vascular endothelial growth factor (VEGF), exhibited a sustained release, within 7 days approximately 60% of the total VEGF [55]. The protein release rate can be controlled by altering the degradation rate of alginate hydrogel. Also, sodium alginate-bacterial cellulose hydrogels had shown promising potential for carrying protein-based drugs, lysozyme (LYZ), via electrostatic adsorption [56].

3.3 Tissue engineering

Tissue engineering, proposed by National Science Foundation in 1987, belongs to the field of biomedical engineering, which is a valuable approach and can be used to restore, maintain, enhance tissues and organs [57]. Tissue engineering involves the getting of seed cells, biological scaffold materials, preparation of tissue and organs, and its clinical application. The biological scaffold materials were utilized to encapsulate and support the propagation of cells. Afterward, these materials were transferred into the body and the target tissues were eventually formed. The research principal field contains bone [58], cartilage [59], vascular [60], ocular tissues [61], skin [62], and other tissues [63].

Alginate, as a most commonly known biomaterial with outstanding properties of scaffold-forming, has been extensively investigated and developed to treat the loss or failure of organs in tissue engineering. It usually combines with other substances to form new alginate derivative materials with the methods of physical or chemical. The different features and functions are obtained, accompanied by improved properties for tissue engineering, such as mechanical strength, cell affinity, and gel-forming ability. Therefore, alginate and its composite have received much attention in tissue engineering.

3.3.1 Bone

Bone has the hierarchical structure forming with 70% of nano-hydroxyapatite (HA, $Ca_{10}(PO_4)_6(OH_2)$) and 30% of collagen by weight [64]. It is a rigid connective organ and plays a major role in the movement of organs, involving affording structural framework, mechanical support and protection, mineral storage, and homeostasis [65]. There are several ways to cause bone defects or fractures in daily life, involving sports injury, accident damage, osteoporosis, osteoarthritis, bone neoplasm, and so on. Bone tissue engineering, as an effective alternative treatment for restoring bone defects, has addressed much attention. Several therapies have been available to treat bone defects or loss, including autograft (bone from patient's own healthy tissue), allograft (bone from the human donor), and xenograft (from other species) [66]. The final goal of bone tissue engineering is to construct bone tissue that should have the same qualities of structurally, physical, and chemical features as natural bone tissue. Alginate has been extensively investigated to synthetic scaffolding materials for bone reconstruction and transfer cells for bone tissue engineering [67].

Sufficient mechanical strength of scaffold in bone tissue engineering is essential to support bone regeneration. To obtain enough mechanical properties, alginate composite scaffolds were prepared mixing other polymers or inorganic components, such as chitosan, collagen, and hydroxyapatite. Chitosan, as a most abundant cationic polysaccharide, is selected to form alginate-chitosan composites. Chitosan/alginate composite scaffolds are extensively investigated for bone tissue engineering. The rigid strength and structural stability are obtained. S. J. Florczyk et al. prepared a chitosan-alginate scaffold with enhanced compressive strength, 0.79 ~ 1.41 MPa [68]. It has the homogeneous pore structure, and the pore size depends on acetic acid and alginate concentration (**Figure 4**). The cell proliferation potential was also improved when the viscosity was below 300 Pas. The greatest defect closure (71.56 ± 19.74%) was observed at 16 weeks [69]. Chitosan/alginate scaffolds present advantages in stimulating osteogenesis and vascularization [70]. Collagen is usually utilized to prepare scaffolds because of its specific properties, such as inducing cell adhesion and degradation [71]. Collagen I, as the essential component of bone tissue's ECM, contributes to migrant and penetration of osteoblasts and vessels [71, 72]. Therefore collagen/alginate hybrid scaffolds have an effect on the osteogenic ability of osteoblasts. It could promote cell spreading, proliferation, and osteogenic differentiation. S. Sotome et al. demonstrated that hydroxyapatite/collagen-alginate could deliver recombinant human bone morphogenetic protein 2 (rh-BMP2) efficiently [73]. There is bone formation within 5 weeks after implantation, accompanied with no obvious deformation. Hydroxyapatite (HA) is the main component of bone. The structure of alginate combined HA composites is similar to the native extracellular matrix of bone.

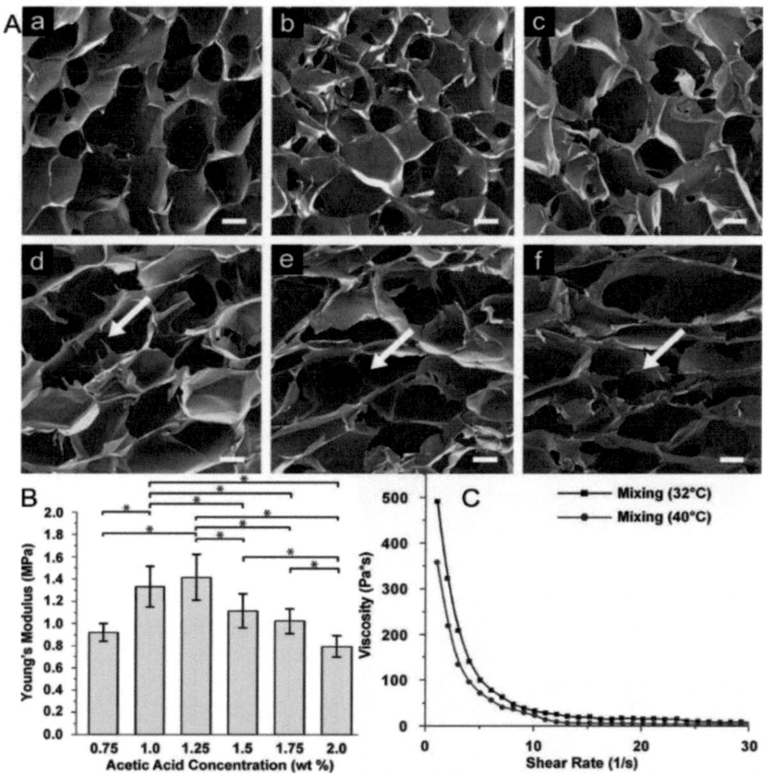

Figure 4.
*(A) SEM images of CA PEC scaffold pore structures made from CA solutions (4 wt % chitosan and 3.75 wt % alginate) with acetic acid concentrations of (a) 0.75 wt %, (b) 1.0 wt %, (c) 1.25 wt %, (d) 1.5 wt %, (e) 1.75 wt %, and (f) 2.0 wt %. The arrows indicate incomplete interconnects and scale bars are 100 μm. (B) Compressive Young's moduli of CA PEC scaffolds prepared from CA solutions (4 wt % chitosan and 3.75 wt % alginate) with varying acetic acid concentrations. *the significant differences between the scaffold groups for the modulus measurements. (C) Influence of mixing temperature on viscosity of CA PEC solution (1.0 wt % acetic acid, 4 wt % chitosan, and 3.75 wt % alginate) at zero shear rate [68].*

The interconnected porous structures and high porosity promote good cell viability, proliferation rate, adhesion, maintenance of osteoblastic phenotype, and bone regeneration [74]. The results of Lin et al. showed that alginate/HA is the better composite porous scaffold with an average pore size of 150 μm and over 82% porosity [75]. In another report, the alginate/HA composite scaffolds have 84% porosity [76]. The uniform pore morphology is favorable to improve compressive strength and elastic modulus, which is proportional to the content of HA. The satisfactory scaffold should have excellent performance to promote bone tissue growth, such as mechanical strength, cell proliferation, and morphology [77]. N. Firouzi et al. found that the addition of HA in alginate-based hydrogel could reduce degradation rate to 41.5%, and significantly improve compressive modulus, reaching 294 ± 2.5 kPa [78]. Meanwhile, microencapsulated osteoblast-like cells showed more proliferation as well as metabolic activities when they were cultured in Alg-Gel Ph-nHA microcapsules during the culture period.

3.3.2 Cartilage

The articular cartilage is an organized and specialized tissue that is highly hydrated (up to 80%), aneural, devoid of blood or lymphatic vessels [79]. It is the connective tissue covering the surface of articulating bones, that provides the lubrication and mechanical strength for body weight and movement. Other body organs are also made by cartilage, such as the ear, nose, bronchial tubes, and so on. Cartilage has a restricted self-repair and regenerative capacity because of lacking nerves and blood vessels. It cannot heal appropriately after the injury, which eventually causes osteoarthritis and cartilage damage. Therefore, repair or reconstruction of damaged cartilage is still a major challenge. Tissue engineering is the potential approach to solve damaged or degraded cartilage. It could mimic the structure and function of body cartilage tissue, stimulating cartilage growth and restoration at the damaged sites. The tissue engineering scaffold has the three-dimensional structure for cell attachment/proliferation.

Alginate scaffolds have been proved to apply for the regeneration of cartilage tissue. It can induce redifferentiation of 2D culture-expanded dedifferentiated chondrocytes [80], therefore alginate could promote the growth of chondrocytes cells, restore damaged cartilage, and keep chondrocyte properties [81]. Chitosan-alginate scaffold effectively promotes the culture of osteogenic and chondrogenic cells. A.E. Erickson et reported that alginate-chitosan + hydroxyapatite scaffolds displayed a defined, interconnected porous network structure [82]. The compressive modulus and stiffness increased with polymer content. After culturing with

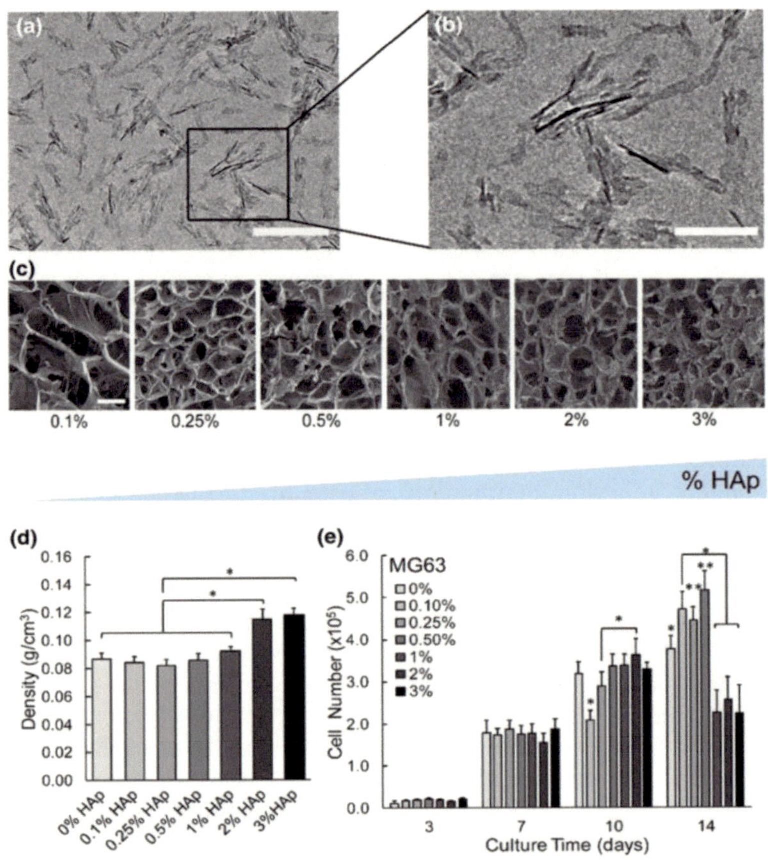

Figure 5.
*Effect of HAp concentration on 6% CA scaffold properties. a, b visualization of HAp nanorods with TEM. Scale bars represent 200 nm, and 100 nm, respectively, in (a) and (b). c SEM images of scaffold pore morphology with varying HAp (HAp concentration increases from left to right). Scale bar represents 200 μm. d density of CA + HAp composite scaffolds (n = 8) ($p \leq 0.05$). (e) MG63 proliferation on CA + HAp composite scaffolds over a 2-week culture period (n = 4). *statistically significant from all or specified conditions ($p \leq 0.05$). **statistically significant from all conditions except 0.1% HAp ($p \leq 0.05$) [80].*

chondrocyte-like (mesenchymal stem cells, MSC), the number of cells on 4% CHA scaffolds was significantly higher than the number of MSCs on 6% CHA scaffolds at day 10 (**Figure 5**) [80]. Hyaluronate, in addition, was explored to reinforce alginate to improve chondrocyte proliferation. It is demonstrated that alginate-hyaluronic acid hydrogel has stable physicochemical properties. C. Mahapatra et reported that alginate-hyaluronic acid-collagen hydrogel provided a binding motif for chondrocytes [83]. These gels effectively protected chondrogenic phenotypes after three weeks of cultivation. The results demonstrated the efficient expression of chondrogenic genes and the formation of cartilage ECMs. Another alginate hybrid gel made combining with polyacrylamide was confirmed to have remarkable mechanical strength, such as high toughness (up to 9000 J/m^2) and stretch ratio [84, 85]. It has

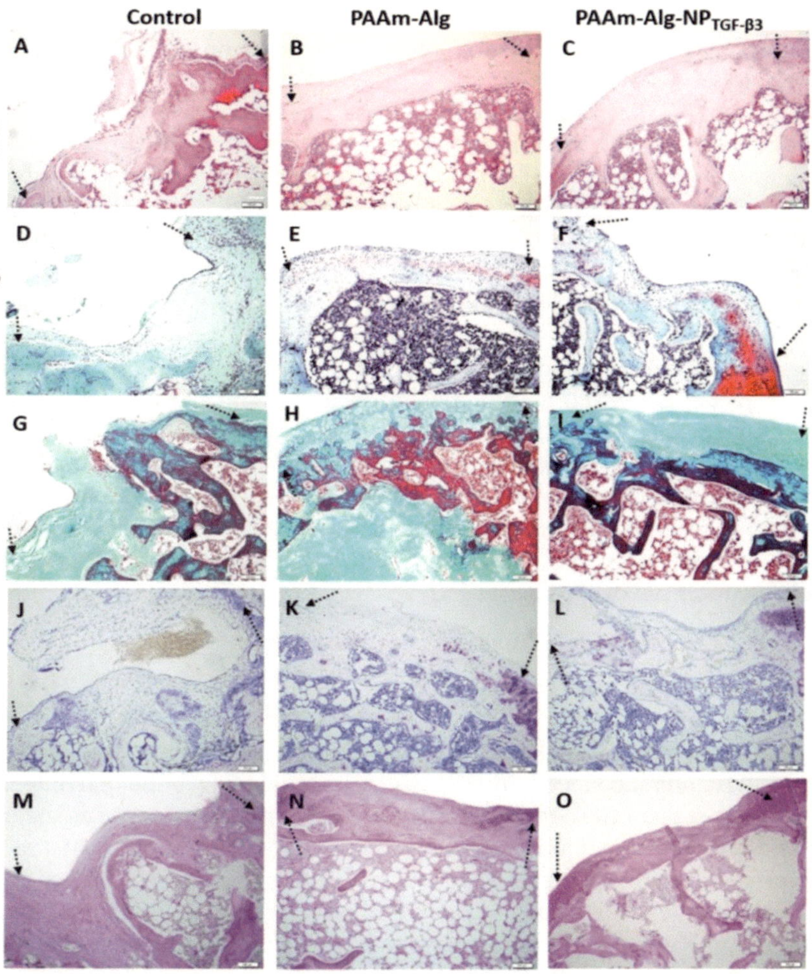

Figure 6.
In vivo representative photomicrographs of tissue sections in the experimental groups stained with (A, B, C) H&E; (D, E, F) safranin O/fast green, (G, H, I) Gomori's trichrome stains, TB (J, K,L), and PAS (M, N,O). (A, D, G) control group demonstrates mostly fibrous tissue with limited tissue repair in the chondral region. (B, E, H) PAAm-Alg group shows fibrous tissue with few fibroblast and chondrocyte-like cells in the joint surface chondral regions. (E) Fibrous tissue in the deeper regions is still observed in some rats. (C, F, I) PAAm-Alg-NP$_{TGF-β3}$ group reveals fibrous tissue with some fibrocartilage and hyaline cartilage groups in the chondral regions. PAAm-Alg and PAAm-Alg-NP$_{TGF-β3}$ groups demonstrated higher (K, L) glycosaminoglycan deposition and (N, O) glycoprotein and proteoglycan contents compared to control group (the arrows indicate the margins of the defect area; scale bars: 100 μm) [87].

also been proved that alginate-polyacrylamide showed good tribological behavior, ultra-low coefficient, and high wear-resistance [86]. After transplantation, chondrocytes displayed an organized distribution and superior integration with surrounding tissue (**Figure 6**) [87]. Alginate and its composites provide chondrogenic differentiation and cell proliferation, therefore, these hybrid gels can be used as cartilage implants.

3.3.3 Liver tissue engineering

Liver disease is a great threat to human health, and it is one of the major reasons for the increased mortality. There are about 2 million deaths all over the world every year [88]. Liver transplantation is the most effective way to solve this problem. Yet, this is a very difficult process because of lacking suitable liver donors. Liver tissue engineering is considered the best way to provide liver to meet the excessive requirement of the liver. It could improve the function of the liver and form a complete organ.

Alginate composites can be used for hepatocyte growth in liver tissue engineering. For example, R. Rajalekshmi et al. reported that fibrin (FIB) incorporated injectable alginate dialdehyde (ADA) - gelatin (G) hydrogel effectively supports growth, proliferation, and functions of hepatic cells [89]. The reason is that fibrin provides several cell adhesion pockets for cell attachment. Liver tissue engineering scaffold, extracellular matrix (ECM), were synthesized with oxidized alginate and galactosylated chitosan via Schiff base reaction. After being cultured in the scaffolds, the hepatocytes exhibited spheroidal morphology. And the multi-cellular aggregates and perfect integration were observed [90]. In addition, the formation of microcapsules is another method for liver tissue engineering. Microcapsules have the ability to encapsulate cells into microbeads. Subsequently, microbeads were covered by the semipermeable membrane to form microcapsules [91]. Microcapsules provide safety microenvironment for cells and protect the cell from interference. After encapsulated HepG2 cells into alginate-based microbeads, the proliferate and protein were clearly observed for at least 12 d [92]. These researches suggested that alginate is a potential candidate for LTE strategies.

4. Conclusion and future prospects

Alginate, as a potential biomaterial, has been successfully explored in different applications such as wound dressing, drug delivery, bone, cartilage. It could afford a moist microenvironment for wound dressing, serve as the carrier for drug delivery, and act as a scaffold for tissue engineering. The outstanding characteristic of alginate for its applications contains biocompatibility, degradable properties, gelatinization capacity, and effective modification to obtain new performances. However, alginate gel suffers from the lack of cell adhesive and mechanical properties, that cause the structural deformation of the scaffold. Despite some strategies that have been carried out to solve these problems, the disadvantages still exist such as lower mechanical properties compared with nature cartilage and lower drug delivery efficiency. For wound dressing, it lacks enough robust and flexible to allow adherence to the skin for a period of time, which maximizing patient uncomfortable and inconvenience. In the future, more novel alginate composites with controlled properties should be constructed by chemical or physical modification. That will play a vital role in intricating drug or cell-loading. Novel alginate composites also could provide mild and targeted degradation properties.

Acknowledgments

This research was supported by the Natural Science Foundation of Jiangsu Province (BK20200791). Oseweuba also gratefully acknowledges the financial support of Wallonia-Brussels International via the Wallonie-Bruxelles International (WBI) excellence Postdoctoral fellowship.

Author details

Yaling Deng[1], Ningning Yang[2], Oseweuba Valentine Okoro[3], Amin Shavandi[3*] and Lei Nie[2*]

1 College of Intelligent Science and Control Engineering, Jinling Institute of Technology, Nanjing, China

2 College of Life Sciences, Xinyang Normal University, Xinyang, China

3 BioMatter Unit—3BIO—École polytechnique de Bruxelles, Université Libre de Bruxelles (ULB), Brussels, Belgium

*Address all correspondence to: amin.shavandi@ulb.be and nieleifu@yahoo.com

References

[1] Smidsrød O, Skja°k-Br⊣k G. Alginate as immobilization matrix for cells. Trends in Biotechnology. 1990;8:71-78.

[2] Kadokawa J-i, Saitou S, Shoda S-i. Preparation of alginate-polymethacrylate hybrid material by radical polymerization of cationic methacrylate monomer in the presence of sodium alginate. Carbohydrate Polymers. 2005;60(2):253-258.

[3] Sachan NK, Pushkar S, Jha A, Bhattcharya AJJPR. Sodium alginate: the wonder polymer for controlled drug delivery. Journal of Pharmacy Research. 2009;2(8):1191-1199.

[4] Sharma A, Gupta MN. Three phase partitioning of carbohydrate polymers: separation and purification of alginates. Carbohydrate Polymers. 2002;48(4):391-395.

[5] Yang J-S, Xie Y-J, He W. Research progress on chemical modification of alginate: A review. Carbohydrate Polymers. 2011;84(1):33-39.

[6] Rhein-Knudsen N, Ale MT, Ajalloueian F, Meyer AS. Characterization of alginates from Ghanaian brown seaweeds: Sargassum spp. and Padina spp. Food Hydrocolloids. 2017;71:236-244.

[7] Otterlei M, Østgaard K, Skjåk-Bræk G, Smidsrød O, Soon-Shiong P, Espevik T. Induction of Cytokine Production from Human Monocytes Stimulated with Alginate. Journal of Immunotherapy. 1991;10(4):286-291.

[8] Johnson FA, Craig DQM, Mercer AD. Characterization of the Block Structure and Molecular Weight of Sodium Alginates. Journal of Pharmacy and Pharmacology. 1997;49(7):639-643.

[9] Fu S, Thacker A, Sperger DM, Boni RL, Buckner IS, Velankar S, et al. Relevance of Rheological Properties of Sodium Alginate in Solution to Calcium Alginate Gel Properties. AAPS PharmSciTech. 2011;12(2):453-460.

[10] D'Ayala GG, Malinconico M, Laurienzo P. Marine Derived Polysaccharides for Biomedical Applications: Chemical Modification Approaches. Molecules. 2008;13(9): 2069-2106.

[11] Shimokawa T, Yoshida S, Takeuchi T, Murata K, Ishii T, Kusakabe I. Preparation of Two Series of Oligo-guluronic Acids from Sodium Alginate by Acid Hydrolysis and Enzymatic Degradation. Bioscience, Biotechnology, and Biochemistry. 1996;60(9): 1532-1534.

[12] Rinaudo MJPB. On the abnormal exponents a ν and a D in Mark Houwink type equations for wormlike chain polysaccharides. Polymer Bulletin. 1992;27(5):585-589.

[13] LeRoux MA, Guilak F, Setton LA. Compressive and shear properties of alginate gel: Effects of sodium ions and alginate concentration. Journal of Biomedical Materials Research. Part B, Applied Biomaterials. 1999;47(1):46-53.

[14] Larsen BE, Bjørnstad J, Pettersen EO, Tønnesen HH, Melvik JE. Rheological characterization of an injectable alginate gel system. BMC Biotechnology. 2015;15(1):29.

[15] Jiménez-Escrig A, Sánchez-Muniz FJ. Dietary fibre from edible seaweeds: Chemical structure, physicochemical properties and effects on cholesterol metabolism. Nutrition Research. 2000;20(4):585-598.

[16] Tam SK, Dusseault J, Bilodeau S, Langlois G, Hallé J-P, Yahia LH. Factors influencing alginate gel biocompatibility. Journal of Biomedical

Materials Research, Part A. 2011;98A(1):40-52.

[17] Zimmermann U, Klöck G, Federlin K, Hannig K, Kowalski M, Bretzel RG, et al. Production of mitogen-contamination free alginates with variable ratios of mannuronic acid to guluronic acid by free flow electrophoresis. Electrophoresis. 1992;13(5):269-274.

[18] de Vos P, Faas MM, Strand B, Calafiore R. Alginate-based microcapsules for immunoisolation of pancreatic islets. Biomaterials. 2006;27(32):5603-5617.

[19] Park H, Park KJPr. Biocompatibility issues of implantable drug delivery systems. Pharmaceutical Research. 1996;13(12):1770-1776.

[20] Tibbitt MW, Anseth KS. Dynamic Microenvironments: The Fourth Dimension. 2012;4(160):160ps24-ps24.

[21] Lueckgen A, Garske DS, Ellinghaus A, Desai RM, Stafford AG, Mooney DJ, et al. Hydrolytically-degradable click-crosslinked alginate hydrogels. Biomaterials. 2018;181:189-198.

[22] Sun J, Tan H. Alginate-Based Biomaterials for Regenerative Medicine Applications. Materials. 2013;6(4): 1285-1309.

[23] Hunt NC, Smith AM, Gbureck U, Shelton RM, Grover LM. Encapsulation of fibroblasts causes accelerated alginate hydrogel degradation. Acta Biomaterialia. 2010;6(9):3649-3656.

[24] Al-Shamkhani A, Duncan R. Radioiodination of Alginate via Covalently-Bound Tyrosinamide Allows Monitoring of its Fate In Vivo. Journal of Bioactive and Compatible Polymers. 1995;10(1):4-13.

[25] Hoque J, Haldar J. Direct Synthesis of Dextran-Based Antibacterial Hydrogels for Extended Release of Biocides and Eradication of Topical Biofilms. ACS Applied Materials & Interfaces. 2017;9(19):15975-15985.

[26] Guest JF, Ayoub N, McIlwraith T, Uchegbu I, Gerrish A, Weidlich D, et al. Health economic burden that different wound types impose on the UK's National Health Service. International wound journal. 2017;14(2):322-330.

[27] Boateng JS, Matthews KH, Stevens HNE, Eccleston GM. Wound Healing Dressings and Drug Delivery Systems: A Review. Journal of Pharmaceutical Sciences. 2008;97(8):2892-2923.

[28] Aderibigbe BA, Buyana B. Alginate in Wound Dressings. Pharmaceutics. 2018;10(2):42.

[29] Jayakumar R, Prabaharan M, Sudheesh Kumar PT, Nair SV, Tamura H. Biomaterials based on chitin and chitosan in wound dressing applications. Biotechnology Advances. 2011;29(3):322-337.

[30] Queen D, Orsted H, Sanada H, Sussman G. A dressing history. International Wound Journal. 2004;1(1):59-77.

[31] Li Y, Han Y, Wang X, Peng J, Xu Y, Chang J. Multifunctional Hydrogels Prepared by Dual Ion Cross-Linking for Chronic Wound Healing. ACS Applied Materials & Interfaces. 2017;9(19):16054-16062.

[32] Thomas A, Harding KG, Moore K. Alginates from wound dressings activate human macrophages to secrete tumour necrosis factor-α. Biomaterials. 2000;21(17):1797-1802.

[33] Wang T, Zhang F, Zhao R, Wang C, Hu K, Sun Y, et al. Polyvinyl Alcohol/ Sodium Alginate Hydrogels Incorporated with Silver Nanoclusters via Green Tea Extract for Antibacterial

Applications. Designed Monomers and Polymers. 2020;23(1):118-133.

[34] Topuz F, Henke A, Richtering W, Groll J. Magnesium ions and alginate do form hydrogels: a rheological study. Soft Matter. 2012;8(18):4877-4881.

[35] Dong Y, Dong W, Cao Y, Han Z, Ding Z. Preparation and catalytic activity of Fe alginate gel beads for oxidative degradation of azo dyes under visible light irradiation. Catalysis Today. 2011;175(1):346-355.

[36] Karri VVSR, Kuppusamy G, Talluri SV, Mannemala SS, Kollipara R, Wadhwani AD, et al. Curcumin loaded chitosan nanoparticles impregnated into collagen-alginate scaffolds for diabetic wound healing. International Journal of Biological Macromolecules. 2016;93:1519-1529.

[37] Orlowska J, Kurczewska U, Derwinska K, Orlowski W, Orszulak-Michalak D. In vitro evaluation of immunogenic properties of active dressings %J Current Issues in Pharmacy and Medical Sciences. 2014;27(1):55-60.

[38] Sudheesh Kumar PT, Raj NM, Praveen G, Chennazhi KP, Nair SV, Jayakumar R. In vitro and in vivo evaluation of microporous chitosan hydrogel/nanofibrin composite bandage for skin tissue regeneration. Tissue engineering Part A. 2013;19(3-4):380-392.

[39] Mndlovu H, du Toit LC, Kumar P, Marimuthu T, Kondiah PPD, Choonara YE, et al. Development of a fluid-absorptive alginate-chitosan bioplatform for potential application as a wound dressing. Carbohydrate Polymers. 2019;222:114988.

[40] Zhao W-Y, Fang Q-Q, Wang X-F, Wang X-W, Zhang T, Shi B-H, et al. Chitosan-calcium alginate dressing promotes wound healing: A preliminary study. Wound Repair and Regeneration. 2020;28(3):326-337.

[41] Tarun K, Gobi N. Calcium alginate/PVA blended nano fibre matrix for wound dressing. Indian Journal of Fibre & Textile Research. 2012, 37:127-132.

[42] Demling RH, Leslie DeSanti MD. The rate of re-epithelialization across meshed skin grafts is increased with exposure to silver. Burns. 2002;28(3):264-266.

[43] Chen K, Wang F, Liu S, Wu X, Xu L, Zhang D. In situ reduction of silver nanoparticles by sodium alginate to obtain silver-loaded composite wound dressing with enhanced mechanical and antimicrobial property. International Journal of Biological Macromolecules. 2020;148:501-509.

[44] Bradford C, Freeman R, Percival SL. In vitro study of sustained antimicrobial activity of a new silver alginate dressing. The journal of the American College of Certified Wound Specialists. 2009;1(4): 117-120.

[45] Wiegand C, Heinze T, Hipler UC. Comparative in vitro study on cytotoxicity, antimicrobial activity, and binding capacity for pathophysiological factors in chronic wounds of alginate and silver-containing alginate. Wound Repair and Regeneration. 2009;17(4): 511-521.

[46] Boontheekul T, Kong H-J, Mooney DJ. Controlling alginate gel degradation utilizing partial oxidation and bimodal molecular weight distribution. Biomaterials. 2005;26(15):2455-2465.

[47] He S, Zhong S, Xu L, Dou Y, Li Z, Qiao F, et al. Sonochemical fabrication of magnetic reduction-responsive alginate-based microcapsules for drug delivery. International Journal of Biological Macromolecules. 2020;155:42-49.

[48] Lengert E, Saveleva M, Verkhovskii R, Abramova A, Goryacheva IJML. Novel formulation of glucocorticoid based on silver alginate

microcapsules for intraarticular drug delivery. Materials Letters. 2021;288: 129339.

[49] Kurczewska J, Pecyna P, Ratajczak M, Gajęcka M, Schroeder G. Halloysite nanotubes as carriers of vancomycin in alginate-based wound dressing. Saudi Pharmaceutical Journal. 2017;25(6):911-920.

[50] Lin N, Gèze A, Wouessidjewe D, Huang J, Dufresne A. Biocompatible Double-Membrane Hydrogels from Cationic Cellulose Nanocrystals and Anionic Alginate as Complexing Drugs Codelivery. ACS Applied Materials & Interfaces. 2016;8(11):6880-6889.

[51] Zhong D, Liu Z, Xie S, Zhang W, Zhang Y, Xue W. Study on poly(D,L-lactic) microspheres embedded in calcium alginate hydrogel beads as dual drug delivery systems. Journal of Applied Polymer Science. 2013;129(2): 767-772.

[52] Rezk AI, Obiweluozor FO, Choukrani G, Park CH, Kim CS. Drug release and kinetic models of anticancer drug (BTZ) from a pH-responsive alginate polydopamine hydrogel: Towards cancer chemotherapy. International Journal of Biological Macromolecules. 2019;141:388-400.

[53] Manatunga DC, de Silva RM, de Silva KMN, de Silva N, Bhandari S, Yap YK, et al. pH responsive controlled release of anti-cancer hydrophobic drugs from sodium alginate and hydroxyapatite bi-coated iron oxide nanoparticles. European Journal of Pharmaceutics and Biopharmaceutics. 2017;117:29-38.

[54] Anitha A, Deepagan VG, Divya Rani VV, Menon D, Nair SV, Jayakumar R. Preparation, characterization, in vitro drug release and biological studies of curcumin loaded dextran sulphate–chitosan nanoparticles. Carbohydrate Polymers. 2011;84(3):1158-1164.

[55] Silva EA, Mooney DJ. Effects of VEGF temporal and spatial presentation on angiogenesis. Biomaterials. 2010;31(6):1235-1241.

[56] Ji L, Zhang F, Zhu L, Jiang J. An in-situ fabrication of bamboo bacterial cellulose/sodium alginate nanocomposite hydrogels as carrier materials for controlled protein drug delivery. International Journal of Biological Macromolecules. 2021;170:459-468.

[57] Griffith LG, Naughton G. Tissue Engineering--Current Challenges and Expanding Opportunities. Science. 2002;295(5557):1009-1014.

[58] Li Z, Du T, Ruan C, Niu X. Bioinspired mineralized collagen scaffolds for bone tissue engineering. Bioactive Materials. 2021;6(5):1491-1511.

[59] Nie X, Chuah YJ, Zhu W, He P, Peck Y, Wang D-A. Decellularized tissue engineered hyaline cartilage graft for articular cartilage repair. Biomaterials. 2020;235:119821.

[60] Zhu J, Chen D, Du J, Chen X, Wang J, Zhang H, et al. Mechanical matching nanofibrous vascular scaffold with effective anticoagulation for vascular tissue engineering. Composites Part B: Engineering. 2020;186:107788.

[61] Sharifi S, Islam MM, Sharifi H, Islam R, Koza D, Reyes-Ortega F, et al. Tuning gelatin-based hydrogel towards bioadhesive ocular tissue engineering applications. Bioactive Materials. 2021;6(11):3947-3961.

[62] Yang W, Xu H, Lan Y, Zhu Q, Liu Y, Huang S, et al. Preparation and characterisation of a novel silk fibroin/hyaluronic acid/sodium alginate scaffold for skin repair. International Journal of Biological Macromolecules. 2019;130:58-67.

[63] Rocca DGD, Willenberg BJ, Qi Y, Simmons CS, Rubiano A, Ferreira LF,

et al. An injectable capillary-like microstructured alginate hydrogel improves left ventricular function after myocardial infarction in rats. International Journal of Cardiology. 2016;220:149-154.

[64] Venkatesan J, Bhatnagar I, Manivasagan P, Kang K-H, Kim S-K. Alginate composites for bone tissue engineering: A review. International Journal of Biological Macromolecules. 2015;72:269-281.

[65] Sowjanya JA, Singh J, Mohita T, Sarvanan S, Moorthi A, Srinivasan N, et al. Biocomposite scaffolds containing chitosan/alginate/nano-silica for bone tissue engineering. Colloids and Surfaces B: Biointerfaces. 2013;109:294-300.

[66] Venkatesan J, Nithya R, Sudha PN, Kim S-K. Chapter Four - Role of Alginate in Bone Tissue Engineering. In: Kim S-K, editor. Advances in Food and Nutrition Research. 73: Academic Press; 2014. p. 45-57.

[67] Giri T, Thakur D, Alexander A, Ajazuddin, Badwaik H, Tripathi DK. Alginate based hydrogel as a potential biopolymeric carrier for drug delivery and cell delivery systems: present status and applications. Current Drug Delivery. 2012;9:539-555.

[68] Florczyk SJ, Kim D-J, Wood DL, Zhang M. Influence of processing parameters on pore structure of 3D porous chitosan–alginate polyelectrolyte complex scaffolds. Journal of Biomedical Materials Research. Part A. 2011;98A(4):614-620.

[69] Florczyk SJ, Leung M, Li Z, Huang JI, Hopper RA, Zhang M. Evaluation of three-dimensional porous chitosan–alginate scaffolds in rat calvarial defects for bone regeneration applications. Journal of Biomedical Materials Research, Part A. 2013;101(10):2974-2983.

[70] Li Z, Ramay HR, Hauch KD, Xiao D, Zhang M. Chitosan–alginate hybrid scaffolds for bone tissue engineering. Biomaterials. 2005;26(18):3919-3928.

[71] Grinnell F. Fibroblast biology in three-dimensional collagen matrices. Trends in Cell Biology. 2003;13(5): 264-269.

[72] García AJ. Get a grip: integrins in cell–biomaterial interactions. Biomaterials. 2005;26(36):7525-7529.

[73] Sotome S, Uemura T, Kikuchi M, Chen J, Itoh S, Tanaka J, et al. Synthesis and in vivo evaluation of a novel hydroxyapatite/collagen–alginate as a bone filler and a drug delivery carrier of bone morphogenetic protein. Materials Science and Engineering: C. 2004;24(3):341-347.

[74] Sharma C, Dinda AK, Potdar PD, Chou C-F, Mishra NC. Fabrication and characterization of novel nano-biocomposite scaffold of chitosan–gelatin–alginate–hydroxyapatite for bone tissue engineering. Materials Science and Engineering: C. 2016;64:416-427.

[75] Lin H-R, Yeh Y-J. Porous alginate/hydroxyapatite composite scaffolds for bone tissue engineering: Preparation, characterization, and in vitro studies. Journal of Biomedical Materials Research, Part B. 2004;71B(1):52-65.

[76] Kim H-L, Jung G-Y, Yoon J-H, Han J-S, Park Y-J, Kim D-G, et al. Preparation and characterization of nano-sized hydroxyapatite/alginate/chitosan composite scaffolds for bone tissue engineering. Materials Science and Engineering: C. 2015;54:20-25.

[77] Petrović M, Čolović B, Jokanović V, Marković DJJoCPR. Self assembly of biomimetic hydroxyapatite on the surface of different polymer thin films. Journal of Ceramic Processing Research. 2012;13(4):398-404.

[78] Firouzi N, Baradar Khoshfetrat A, Kazemi D. Enzymatically gellable

gelatin improves nano-hydroxyapatite-alginate microcapsule characteristics for modular bone tissue formation. Journal of Biomedical Materials Research, Part A. 2020;108(2):340-350.

[79] Armiento AR, Stoddart MJ, Alini M, Eglin D. Biomaterials for articular cartilage tissue engineering: Learning from biology. Acta Biomaterialia. 2018;65:1-20.

[80] Sahoo DR, Biswal T. Alginate and its application to tissue engineering. SN Applied Sciences. 2021;3(1):30.

[81] Bouhadir KH, Lee KY, Alsberg E, Damm KL, Anderson KW, Mooney DJJBp. Degradation of partially oxidized alginate and its potential application for tissue engineering. Biotechnology Progress. 2001;17(5):945-950.

[82] Erickson AE, Sun J, Lan Levengood SK, Swanson S, Chang F-C, Tsao CT, et al. Chitosan-based composite bilayer scaffold as an in vitro osteochondral defect regeneration model. Biomedical Microdevices. 2019;21(2):34.

[83] Mahapatra C, Jin G-Z, Kim H-W. Alginate-hyaluronic acid-collagen composite hydrogel favorable for the culture of chondrocytes and their phenotype maintenance. Tissue Engineering and Regenerative Medicine. 2016;13(5):538-546.

[84] Darnell MC, Sun J-Y, Mehta M, Johnson C, Arany PR, Suo Z, et al. Performance and biocompatibility of extremely tough alginate/polyacrylamide hydrogels. Biomaterials. 2013;34(33):8042-8048.

[85] Liao I-C, Moutos FT, Estes BT, Zhao X, Guilak F. Composite Three-Dimensional Woven Scaffolds with Interpenetrating Network Hydrogels to Create Functional Synthetic Articular Cartilage. Advanced Functional Materials. 2013;23(47):5833-5839.

[86] Arjmandi M, Ramezani M. Mechanical and tribological assessment of silica nanoparticle-alginate-polyacrylamide nanocomposite hydrogels as a cartilage replacement. Journal of the Mechanical Behavior of Biomedical Materials. 2019;95:196-204.

[87] Saygili E, Kaya E, Ilhan-Ayisigi E, Saglam-Metiner P, Alarcin E, Kazan A, et al. An alginate-poly(acrylamide) hydrogel with TGF-β3 loaded nanoparticles for cartilage repair: Biodegradability, biocompatibility and protein adsorption. International Journal of Biological Macromolecules. 2021;172:381-393.

[88] Asrani SK, Devarbhavi H, Eaton J, Kamath PS. Burden of liver diseases in the world. Journal of Hepatology. 2019;70(1):151-171.

[89] Rajalekshmi R, Kaladevi Shaji A, Joseph R, Bhatt A. Scaffold for liver tissue engineering: Exploring the potential of fibrin incorporated alginate dialdehyde–gelatin hydrogel. International Journal of Biological Macromolecules. 2021;166:999-1008.

[90] Chen F, Tian M, Zhang D, Wang J, Wang Q, Yu X, et al. Preparation and characterization of oxidized alginate covalently cross-linked galactosylated chitosan scaffold for liver tissue engineering. Materials Science and Engineering: C. 2012;32(2):310-320.

[91] Majewski RL, Zhang W, Ma X, Cui Z, Ren W, Markel DCJJoab, et al. Bioencapsulation technologies in tissue engineering. Journal of Applied Biomaterials & Functional Materials. 2016;14(4):395-403.

[92] Haque T, Chen H, Ouyang W, Martoni C, Lawuyi B, Urbanska AM, et al. In vitro study of alginate–chitosan microcapsules: an alternative to liver cell transplants for the treatment of liver failure. Biotechnology Letters. 2005;27(5):317-322.

Chapter 7

Applications of Alginates in the Design and Preparation of Orodispersible Dosage Forms

Garba M. Khalid and Francesca Selmin

Abstract

Orodispersible dosage forms are attractive and innovative drug delivery systems that can fulfill individual patient needs, especially in children, elderly and among dysphagic patients. Indeed, they rapidly disperse in the mouth upon contact with the saliva without the need for water or munching. Examples of such dosage forms include orodispersible tablets (ODT), and orodispersible films (ODF). The ability to obtain ODF with different dimensions (sizes and thicknesses) makes them a suitable for personalized dosing of single or a fixed-dose combination of drugs in special patient populations. Several biopolymers are currently being exploited in the development of orodispersible dosage forms including alginates due to their versatility, availability, naturally occurring, and biosafety profile. This chapter provides an appraisal on the various applications of alginates in the preparations and their role on the properties of orodispersible dosage forms and highlights future perspectives of this very versatile biopolymer for these innovative drug delivery systems.

Keywords: alginates, orodispersible films, orodispersible tablets, solid dosage forms, personalized therapy

1. Introduction

Orodispersible dosage forms can ameliorate the lack of compliance associated with the administration of conventional oral solid-dosage forms (i.e., capsules and tablets), or even oral liquid-dosage forms in some patients with swallowing difficulties and hence, have the potential to improve medication adherence [1]. Indeed, since their appearance into the pharmaceutical market, their development has grown gradually, moving from orodispersible tablets (ODT) to orodispersible films (ODF) which presents several advantages to completely eliminate the fear of chocking in some patients [2, 3]. Moreover, pharmaceutical companies have amplified research in these dosage forms because they can easily extend their product portfolio [4].

ODF have the size of a postage stamp and are individually packed so that transportation and patient handling are friendly (**Figure 1a**). ODF consist of a single or multilayer sheet of suitable materials intended to be place in the mouth where they rapidly dispersed upon contact with the saliva without need of water or munching (**Figure 1b**). They provide the opportunity to meet the needs of specific

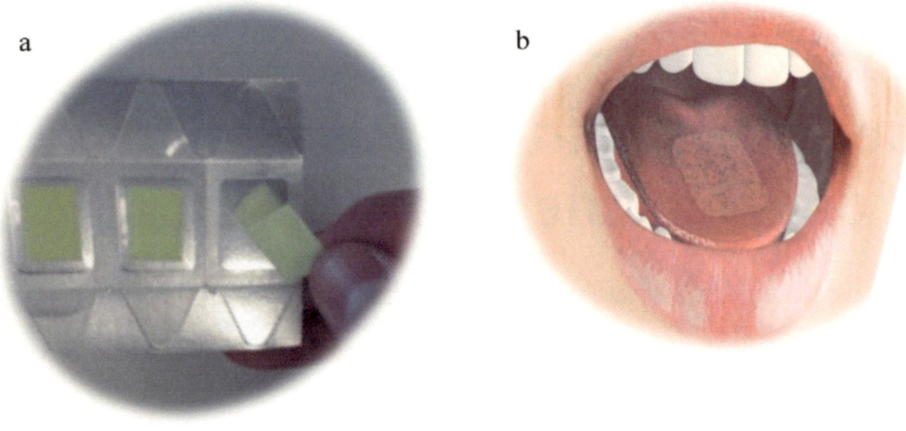

Figure 1.
Typical ODF handling from packaging material (a); ideal ODF administration without water (b).

subpopulation of patients suffering from a variety of disorders such as dysphagia due to pathological or psychological issues. In addition, children and elderly, and patients with limited access to water and/or restricted water intake can also benefit from their merits [1, 5]. Indeed, the possibility to change size, shape and color of the ODF have open new scenarios to prepare small batches for personalization of dose in special patient population [6]. Furthermore, ODF can be advantageously used as a carrier for other technologies, such as microparticles, nanocrystals and self-emulsifying systems [5, 7–9], which regulate the drug release patterns and, hence, its bioavailability. However, the main ODF pitfalls are related to the limited formulation space [10] which implies a limited drug loading capacity. Secondly, palatability drives the compliance for ODF loaded formulations, but the formulation space often limits the addition of taste masking agents; even if both bitter and/or astringent taste of a drug can be opportunely reduced and/or eliminated [11, 12]. Thirdly, the manufacturing process at the industrial scale is mainly based on solvent-casting technologies, which require production chains with specialized equipment common only to transdermal patches, and therefore, the number of manufacturers worldwide are limited. Nevertheless, similar to transdermal patches, the dose loaded in an ODF is defined by their size and, therefore, the same production chain could be used to prepare batches of different drug strengths. Because of this peculiarity, researchers are striving to optimize and/or to develop technologies to exploit this peculiarity in the extemporaneous compounding of small batches of ODF in a pharmacy setting [5]. Since the term "customized dosage form" should be related not only to a tailored dose but also to doses on-demand, shape and color of a dosage form [13], this innovation would also allow end-users to easily identify their own medicine, improving medication safety and adherence [5].

ODF are generally made up of plasticized hydrocolloids or blends made thereof that can be laminated by several techniques and sealed in a moisture-protective packages [1]. The active pharmaceutical ingredient (API) can be dissolved or dispersed as such or as nanocrystals [7] or loaded into microparticles [8] depending on the physicochemical properties of the drug and the desired release pattern. Other ODF formulation components are; surfactants, viscosity modifiers, taste-masking agents and coloring agents, when required [5, 14, 15].

Among the critical quality attributes of ODF, satisfactory tensile properties to guarantee packaging and handling during administration without breakage, the disintegration and dissolution in the oral cavity, acceptable taste [5], esthetic appearance, and stability of the dosage form itself and the loaded drug(s) need to be carefully studied. For instance, the choice of taste-masking agents depends not only on the improvement of palatability, but also on their compatibility with other formulation components, the possible impact on the drug's solubility and dissolution rate, and mechanical properties of the final ODF formulation [5, 12]. This chapter provides an appraisal of the various applications of alginates in the design and preparation of orodispersible dosage forms as new emerging drug delivery systems to overcome some limitations with the conventional solid dosage forms. The literature was generated from the Scopus, and PubMed data bases by searching single or the combination of the following keywords; alginate, alginates, orodispersible film, orodispersible tablet, and orodispersible dosage forms.

2. Preparation methods orodispersible dosage forms

Orodispersible dosage forms can be broadly divided into two based on their current commercial availability, i.e., orodispersible tablet (ODT) and orodispersible films (ODF). Each of them requires a specialized type of equipment, polymers and other formulation additives or excipients since they both have some similarity and distinctive peculiarities. The market penetration and commercial success of either ODT or ODF depends on the taste of the finished product since the palatability of the drug product in orodispersible dosage forms determines patient acceptability and subsequently therapeutic success of drug loaded. Therefore, different taste masking strategies are employed to mask unpleasant tastes or odors of drugs in orodispersible dosage forms such as the use of microencapsulation, complexation technique, using taste masking agents such as sweeteners and flavoring agents [3, 14].

Orodispersible tablets (ODT) have the same appearance with the conventional tablets. However, unlike the conventional tablets, they are expected to rapidly disintegrate within 3 min as a result of their high porous network with the rapid penetration of water and/or other fluids. They can be prepared by direct compression, heat molding technique, or freeze-drying using specific excipients such as alginates (**Table 1**) or their particle engineering products [3].

Orodispersible films (ODF) are mainly prepared by solvent casting technology, electrospinning, hot-melt extrusion and more recently by various printing techniques for dose personalization [3, 5]. In some cases, a combination of these methods is used to obtain an ODF with desired properties or to achieve the desired technological and therapeutic objectives. For instance, various printing technologies have been coupled with other ODF preparation techniques such as solvent casting and fused deposition modeling (FDM) 3D printing to prepared ODF on-demand. Moreover, hot-melt extrusion (HME) is a solvent-free, continuous process. It has a short processing time, suitable for small-scale on-demand preparation of medicines, and is easy to scale-up. Several thermoplastic polymer-carriers and other additives used during extrusion processing are generally regarded to be safe for human consumption [28]. Over the last two decades, HME has been employed as a novel cost-effective pharmaceutical manufacturing technique of different oral solid-dosage forms. It has been suitably used in the preparation of immediate-release, novel taste-masked and abuse deterrence tablets formulations [28, 29], chrono-modulated drug delivery systems [30], for immediate release formulation of ODF [2, 3], and ODF containing poorly water soluble and highly polymorphic drugs [31–33]. Thus, the combination of HME and additive printing technology has been shown to offer several advantages [6, 33]; first, the ability to fabricate immediate-release, modified-release, and other novel drug delivery dosage forms, second, the ease to prepare personalized oral drug delivery products,

Type of orodispersible dosage form	Type of alginate used	Role of the alginate in the formulation	Special notes	Reference
ODF	Sodium alginate and alginate oligosaccharides	Drug carriers	• Both placebo and posaconazole mucoadhesive films formulations containing the sodium alginate and alginate oligosaccharides exhibited antifungal properties on Candida species.	[16]
ODF	Alginate hexyl amide derivative	Film-forming polymer	• New film-forming alginate hexyl amide derivative was prepared. • Repaglinide bioavailability was enhanced by the new film-forming alginate derivative.	[17]
ODF	Sodium alginate	Film-forming polymer	• Nebivolol hydrochloride was loaded in the ODF using solvent casting technique. • Croscaramellose sodium was used as super-disintegrant in the formulation	[18]
ODF	Sodium alginate	Film-forming polymer	• Fluconazole mucoadhesive films were prepared to provide prolonged release of the drug for topical treatment of candidiasis. • Sodium alginate was used alone or in combination with sodium carboxymethyl cellulose, Carbopol, and polycarbophil.	[19]
ODF	Sodium alginate	Film-forming polymer	• Terbutaline sulphate sublingual films were prepared by solvent casting with the sodium alginate singly or in combination with xanthan gum, Carbopol, or HPMC E5 and maltodextrin. • The films disintegrate rapidly with faster absorption rate compared to conventional tablet from the pharmacokinetic study on healthy human volunteers. • The formulated films can be use in the management of acute episodes of asthma attacks.	[20]
ODF	Sodium alginate	Film-forming polymer	• Levocetirizine HCl film was prepared with the addition of sodium starch glycolate as disintegrating agent. • A full 3^2 factorial design was applied to optimize the concentration of the sodium alginate and disintegrant	[21]
ODF	Sodium alginate	Film-forming polymer	• Piroxicam ODF were prepared by solvent casting.	[22]
ODF	Sodium alginate	Disintegrant	• Sildenafil citrate was loaded in the ODF resulted in decreased disintegration time with increasing concentration of the sodium alginate in the formulation.	[23]

DOI: http://dx.doi.org/10.5772/intechopen.98610

Type of orodispersible dosage form	Type of alginate used	Role of the alginate in the formulation	Special notes	Reference
ODT	Calcium alginate	Superdisintegrant	• Mini ODT were prepared by direct compression using mannitol and MCC as filler/ binder excipients. • Addition of the calcium alginate have shown to improve the disintegration time of the mini ODT.	[24]
ODT	Alginic acid and calcium alginate	Superdisintegrant	• Xerogels containing alginic acid, or calcium alginate and a mixture of both were isolated using oven and rotary evaporation methods. • Mini ODT were prepared with the xerogels developed from the alginates with improved wettability and ultimately rapid disintegration.	[25]
ODT	Sodium alginate	Taste masking agent	• Solid dispersion was used to deter the taste of fexofenadine HCl by controlling the rate of drug release in the saliva pH	[26]
ODT	Sodium alginate	Superdisintegrant	• Meclizine HCl ODT was formulated by direct compression using sodium alginate as such or in combination with xanthan gum to reenforce the superdisintegrant properties of the former and to improve the mechanical quality of the tablets.	[27]

HPMC: hydroxypropyl methyl cellulose; MCC: microcrystalline cellulose; HCl: hydrochloride.

Table 1.
Applications of alginates in orodispersible dosage forms.

thirdly, the ability to load drugs that are prone to polymorphic transformation in the present of water or solvent system [33], and finally, the ability to reduces several unit-operations related to traditional manufacturing process such as granulation, milling, sieving, compressing, and coating. Hence, they are very efficient and economical, in particular for small-scale on-demand personalized formulations [34, 35]. Moreover, additive printing technology combines digital design, manufacturing, and controls together, which is an accurate, timesaving, continuous process to meet individual patient needs [28, 36]. Therefore, the combination of HME and additive printing can be applied as a fabrication tool in the field of digital health for the remote manufacture and dispensing of personalized orodispersible dosage forms having shapes, sizes, and doses that are optimized for a particular patient or group of patients based on their needs [28]. Consequently, the therapeutic efficacy and medication adherence can be enhanced owing to the flexibility and autonomy of the treatment process provided by the combination of these technologies.

3. Application of additive manufacturing in personalized orodispersible dosage forms

The concept of personalized therapy is gaining attention in recent years with the aim of providing appropriate drug(s), dosage regiments, and/or medical care based on the individual patient's peculiarities such as age, medical history, diagnostic results, or genetic information aiming to minimize medication errors such as drug interactions, side effects, and adverse drug reactions and to obtain maximum therapeutic benefits [37, 38]. Technologically speaking, personalized therapy is centered towards personalized dosing, or dose precision and providing age-appropriate dosage forms capable of addressing patients' specific requirements. As an example, in the case of children, it has been estimated that the availability of authorized and commercially available medicine for children varies between 48% and 54% of all approved medicines and that up to 50% of pediatric patients receive an unlicensed or off-label prescription [39]. Indeed, it has been recognized that children are unable to or have difficulties with swallowing tablets or capsules. Moreover, crushing tablets, opening capsules, or mixing powders to extemporaneously prepare the required dose with liquids may lead to dose variability, contamination, drug instability, taste and solubility problems, and other consequences for safety of the patient and the efficacy of the treatment. It is therefore of paramount importance to develop age-appropriate dosage forms, as pointed out by the World Health Organization's (WHO) initiative 'better medicines for children'. Therefore, personalized orodispersible dosage forms such as ODF or ODT prepared on-demand by additive printing technologies can potentially legitimized personalized and precise dosing in patients with special needs.

Additive printing presents a promising future for the point-of-care manufacturing of medicines. Technologies such as two-dimensional (2D) and three-dimensional (3D) printing appeared capable of producing individualized oral drug delivery systems, such as ODF with customizable drug dose strengths, and in some cases with pre-defined drug release patterns [40, 41]. The use of computer-aided design (CAD) software to design different dimensions of a 3D objects prior to printing displays the potential of 3D printing for the concept of precision dosing of active pharmaceutical ingredients (APIs) [41]. Despite the significant technological advancements made so far during the 21st century on conventional pharmaceutical manufacturing processes, especially being cost-effective for large-scale industrial production, they can be inherently labour intensive, time-consuming, and dose inflexible. This poses significant challenges for certain groups of patients that require tailored dosing (particularly among pediatrics and geriatrics) or for certain medicines that require frequent dose adjustments (e.g., drugs

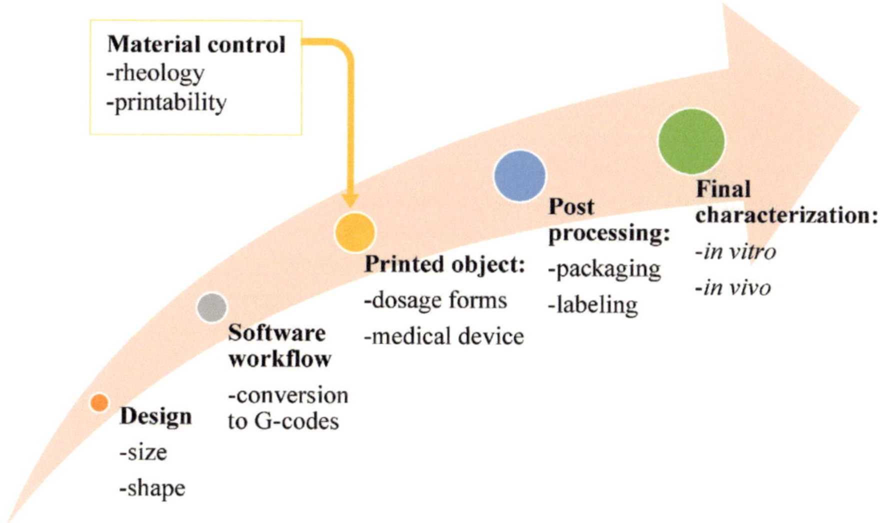

Figure 2.
Schematic flow chart of additive printing process from design and material selection to the finished product characterization.

with narrow therapeutic index). Therefore, to achieve pharmacotherapeutic goals with greater efficacy, quality and safety in patients with special needs, the use of innovative approach such as additive printing technologies are required within pharmaceutical field to facilitate the preparation of small-scale, on-demand and dose-flexible formulations such as ODF. Additive printing process which enables the design of a customized oral dosage forms is triggering a paradigm shift in the way medicines are manufactured and administered [42]. Furthermore, this process could make possible the printing of medicines in pandemic outbreak areas to mitigate drug shortages and supply chain disruptions, and potential for making available printing of medicines in war zones, in clinical trials in hospital settings [43, 44] and preparation of individualized fixed-dose combination products [44–46]. Perhaps, the current regulatory landscape on additive printing is flexible enough to accommodate this technology for mass production in addition to its benefits in extemporaneous compounding of medicines. Thus, according to United States Food and Drug Administration's (FDA) guidelines for additive printing, once the printing device/equipment is optimized, the first step is the design process, which can include a standard design with distinct pre-specified dimensions, in the case of ODF for instance; the design of an ODF area and desired thickness which ultimately defined the dose of loaded API according to individual patient needs. Once the device design is converted to a digital file, the software workflow phase begins, and that file is further processed to prepare it for printing (**Figure 2**), at this stage, the printing parameters are optimized. Concurrently with this step, material controls are established for materials used in the printing of the dosage form (i.e., rheological evaluation, and printability). After printing is complete, post-processing of the built dosage form (e.g., packing and labelling) takes place. After post-processing, the final finished dosage form is ready for characterization [47].

4. Applications of alginates in orodispersible dosage forms

Alginate is a natural polymer used widely in pharmaceutical, food, and biomedical applications because of its biodegradable and biocompatible properties. Alginate and its derivatives are considered low or nontoxic, and non-immunogenic hence suitable for human consumption [48, 49]. Moreover, various alginate salts (ammonium, calcium, sodium, potassium) and propylene glycol alginate derivatives are generally regarded as safe (GRAS) ingredients, for oral administration by the FDA. Like many other pharmaceutical products, the selection of excipients for orodispersible dosage forms requires thoughtful consideration based on the peculiar properties of the dosage form itself such as rapid disintegration, drug-excipient compatibility, biocompatibility with body fluids such as saliva and GIT fluids. Therefore, alginate and its derivatives have found different applications in the design and preparation of orodispersible dosage forms. As an example, sodium alginate has been used as an orodispersible film-forming polymer to provide mucoadhesive properties of the films or to increased drug loading or both. These applications are mostly accomplished with the sodium alginate alone to load poorly soluble drugs in ODF [22], or in combination with other polymers such as polyvinyl alcohol, chitosan, Carbopol 974P, and sodium carboxymethyl cellulose [22, 23, 50]. Various ODF formulations applying the uniqueness of alginates to provide different functions in the dosage form are exemplified in (**Table 1**). Indeed, it is worthy to note that, alginate-based orodispersible films have shown high tensile strength and high hydrophilicity with disintegration time within few seconds [20, 51]. Hence, the use of alginate in ODF represent good systems to formulate orodispersible dosage forms, promoting hydrophilic properties and ultimately suitable disintegration time which is fundamental quality for ODF. Nevertheless, few studies have explored alginate as a film-forming agent in ODF preparations either singly or in combination with other polymers. Thus, highlighting missing gaps to explore in discovering other potentials of alginates in ODF formulations [51].

Similarly, various alginate derivatives have also found application in the formulation of ODT where they serve different functionality. For instance, alginic acid and its salts have been used as disintegrants, super-disintegrants, and/or fillers in ODT formulations [24, 25]. Soulairol and co-workers developed xerogels containing alginic acid, calcium alginate or the combination thereof in the design and formulation of orodispersible mini tablets where they have been shown to provide enhanced super-disintegrant properties [25]. Further, Yehia et al., employed the use of sodium alginate as a taste-masking agent in ODT through solid dispersion technique. The authors succeeded in masking the taste of fexofenadine hydrochloride by controlling the rate of drug release in the saliva pH as confirmed by healthy human volunteers [26]. A summary of the various applications of alginates in ODT is also provided in (**Table 1**).

5. Conclusion

Alginates have wide range of application in the design and preparation of orodispersible dosage forms. The choice of a particular alginate type depends on several factors, such as the type of the orodispersible dosage form to be prepared, i.e., ODT or ODF, the physicochemical properties of the alginate and that of the drug itself, and the nature of the release pattern of the loaded drug(s). Most of the alginate's applications in ODF and ODT are in the solvent casting and direct compression method for their preparations, respectively. Thus, exploring alginates applications in the preparation of these dosage forms by other techniques such as 3D printing, freeze drying, and particle engineering can open new scenarios in their application in the development of these innovative drug delivery systems.

Author details

Garba M. Khalid[1*] and Francesca Selmin[2]

1 Department of Pharmaceutics and Pharmaceutical Technology, Bayero University, Kano, Nigeria

2 Department of Pharmaceutical Sciences, Universita degli Studi di Milano, Milan, Italy

*Address all correspondence to: khalidgmk@gmail.com

References

[1] Cilurzo F, Musazzi UM, Franzé S, et al. Orodispersible dosage forms: biopharmaceutical improvements and regulatory requirements. Drug Discov. Today [Internet]. 2018;23:251-259. Available from: http://dx.doi.org/10.1016/j.drudis.2017.10.003.

[2] Manjeet B. Pimparade, Anh Vo, Abhijeet S. Maurya, Jungeun Bae, Joseph T. Morott, Xin Feng, Dong Wuk Kim, Vijay I. Kulkarni, Roshan Tiwari, K. Vanaja, Reena. Murthy, H. N. Shivakumar, D. Neupane, S. R. Mishra, S. N. Murthy and MAR. Development and Evaluation of an Oral Fast Disintegrating Anti-allergic Film Using Hot-melt Extrusion Technology. Eur. J. Pharm. Biopharm. [Internet]. 2017;119:81-90. Available from: http://dx.doi.org/10.1016/j.ejpb.2017.06.004.

[3] Cilurzo F, Musazzi UM, Franzé S, et al. Orodispersible dosage forms: biopharmaceutical improvements and regulatory requirements. Drug Discov. Today. 2018;23.

[4] Garba M. Khalid, Francesca Selmin, Umberto M. Musazzi, Chiara GM Gennari PM and FC. Trends in the Characterization Methods of Orodispersible Films. Curr. Drug Deliv. 2020;1-16.

[5] Musazzi UM, Khalid GM, Selmin F, et al. Trends in the production methods of orodispersible films. Int. J. Pharm. 2020;576.

[6] Khalid GM, Musazzi UM, Selmin F, et al. Extemporaneous printing of diclofenac orodispersible films for pediatrics Extemporaneous printing of diclofenac orodispersible films for pediatrics. Drug Dev. Ind. Pharm. [Internet]. 2021;0:1-9. Available from: https://doi.org/10.1080/03639045.2021.1908335.

[7] Lai F, Franceschini I, Corrias F, et al. Maltodextrin fast dissolving films for quercetin nanocrystal delivery. A feasibility study. Carbohydr. Polym. [Internet]. 2015;121:217-223. Available from: http://dx.doi.org/10.1016/j.carbpol.2014.11.070.

[8] Musazzi UM, Passerini N, Albertini B, et al. A new melatonin oral delivery platform based on orodispersible films containing solid lipid microparticles. Int. J. Pharm. [Internet]. 2019;559:280-288. Available from: https://doi.org/10.1016/j.ijpharm.2019.01.046.

[9] Talekar SD, Haware R V., Dave RH. Evaluation of self-nanoemulsifying drug delivery systems using multivariate methods to optimize permeability of captopril oral films. Eur. J. Pharm. Sci. [Internet]. 2019;130:215-224. Available from: https://doi.org/10.1016/j.ejps.2019.01.039.

[10] Borges AF, Silva C, Coelho JFJ, et al. Outlining critical quality attributes (CQAs) as guidance for the development of orodispersible films. Pharm. Dev. Technol. 2017;22:237-245.

[11] Cilurzo F, Cupone IE, Minghetti P, et al. Nicotine Fast Dissolving Films Made of Maltodextrins: A Feasibility Study. AAPS PharmSciTech [Internet]. 2010;11:1511-1517. Available from: http://www.springerlink.com/index/10.1208/s12249-010-9525-6.

[12] Cilurzo F, Cupone IE, Minghetti P, et al. Diclofenac fast-dissolving film: Suppression of bitterness by a taste-sensing system. Drug Dev. Ind. Pharm. 2011;37:252-259.

[13] Vakili H, Nyman JO, Genina N, et al. Application of a colorimetric technique in quality control for printed pediatric orodispersible drug delivery systems containing propranolol hydrochloride. Int. J. Pharm. [Internet]. 2016;511:606-618. Available from: http://dx.doi.org/10.1016/j.ijpharm.2016.07.032.

[14] Dixit RP, Puthli SP. Oral strip technology: Overview and future potential. J. Control. Release. 2009;139:94-107.

[15] Musazzi UM, Selmin F, Ortenzi MA, et al. Personalized orodispersible films by hot melt ram extrusion 3D printing. Int. J. Pharm. 2018;551:52-59.

[16] Szekalska M, Wróblewska M, Trofimiuk M, et al. Alginate oligosaccharides affect mechanical properties and antifungal activity of alginate buccal films with posaconazole. Mar. Drugs. 2019;17.

[17] Khames A. Hexyl alginate derivative, an amphiphilic innovative buccal film-forming material of promising mechanical and release characteristics for the improvement of repaglinide bioavailability. Drug Des. Devel. Ther. 2019;13:925-940.

[18] Vijayanand P, Patil JS, Reddy MV. Formulation and comparative pharmacokinetic evaluation of orodispersible tablets and films of nebivolol hydrochloride. J. Pharm. Investig. 2015;45:237-247.

[19] Yehia SA, El-Gazayerly ON, Basalious EB. Fluconazole mucoadhesive buccal films: in vitro/in vivo performance. Curr. Drug Deliv. [Internet]. 2009;6:17-27. Available from: http://www.ncbi.nlm.nih.gov/pubmed/19418952.

[20] Sayed S, Ibrahim HK, Mohamed MI, et al. Fast-dissolving sublingual films of terbutaline sulfate: Formulation and in vitro/in vivo evaluation. Mol. Pharm. 2013;10:2942-2947.

[21] Singh S, Satyam G, Garima G, et al. Formulation and Evaluation of Rapidly Disintegrating Film of Levocetrizine Hydrochloride. Der Pharm. Lett. 2010;2:434-439.

[22] El-Bary AA, Al Sharabi I, Haza'a BS. Effect of casting solvent, film-forming agent and solubilizer on orodispersible films of a polymorphic poorly soluble drug: an in vitro/in silico study. Drug Dev. Ind. Pharm. [Internet]. 2019;45:1751-1769. Available from: https://doi.org/10.1080/03639045.2019.1656733.

[23] Shi LL, Xu WJ, Cao QR, et al. Preparation, characterization and in vitro evaluation of a polyvinyl alcohol/sodium alginate based orodispersible film containing sildenafil citrate. Pharmazie. 2014;69:327-334.

[24] Soulairol I, Chaheen M, Tarlier N, et al. Evaluation of disintegrants functionality for orodispersible mini tablets. Drug Dev. Ind. Pharm. 2017;43:1770-1779.

[25] Soulairol I, Sanchez-Ballester NM, Aubert A, et al. Evaluation of the super disintegrant functionnalities of alginic acid and calcium alginate for the design of orodispersible mini tablets. Carbohydr. Polym. [Internet]. 2018;197:576-585. Available from: https://doi.org/10.1016/j.carbpol.2018.06.002.

[26] Yehia SA, El-Ridi MS, Tadros MI, et al. Phenylalanine-free taste-masked orodispersible tablets of fexofenadine hydrochloride: Development, in vitro evaluation and in vivo estimation of the drug pharmacokinetics in healthy human volunteers. Pharm. Dev. Technol. 2015;20:528-539.

[27] Kumar MV. Formulation and Evaluation of Meclizine Hcl Orally Dispersible Tablets by using Natural Super Disintegrants. Int. J. Pharma Sci. Sci. Res. 2016;2:53-80.

[28] Wang H, Dumpa N, Bandari S, et al. Fabrication of Taste-Masked Donut-Shaped Tablets Via Fused Filament Fabrication 3D Printing Paired with Hot-Melt Extrusion Techniques. AAPS PharmSciTech. 2020;21:1-11.

[29] Nukala PK, Palekar S, Patki M, et al. Abuse Deterrent Immediate Release Egg-Shaped Tablet (Egglets) Using 3D Printing Technology: Quality by Design to Optimize Drug Release and Extraction. AAPS PharmSciTech. 2019;20.

[30] Dumpa NR, Sarabu S, Bandari S, et al. Chronotherapeutic Drug Delivery of Ketoprofen and Ibuprofen for Improved Treatment of Early Morning Stiffness in Arthritis Using Hot-Melt Extrusion Technology. AAPS PharmSciTech. 2018;19:2700-2709.

[31] Cho HW, Baek SH, Lee BJ, et al. Orodispersible polymer films with the poorly water-soluble drug, olanzapine: Hot-melt pneumatic extrusion for single-process 3D printing. Pharmaceutics. 2020;12:1-16.

[32] Modica de Mohac L, de Fátima Pina M, Raimi-Abraham BT. Solid microcrystalline dispersion films as a new strategy to improve the dissolution rate of poorly water soluble drugs: A case study using olanzapine. Int. J. Pharm. [Internet]. 2016;508:42-50. Available from: http://dx.doi.org/10.1016/j.ijpharm.2016.05.012.

[33] Selmin F, Khalid GM, Musazzi UM, et al. Relevance of production method on the physical stability and in vitro biopharmaceutical performances of olanzapine orodispersible film. Int. J. Pharm. [Internet]. 2021;120697. Available from: https://doi.org/10.1016/j.ijpharm.2021.120697.

[34] Zhang J, Xu P, Vo AQ, et al. Development and evaluation of pharmaceutical 3D printability for hot melt extruded cellulose-based filaments. J. Drug Deliv. Sci. Technol. [Internet]. 2019;52:292-302. Available from: https://doi.org/10.1016/j.jddst.2019.04.043.

[35] Zhang J, Feng X, Patil H, et al. Coupling 3D printing with hot-melt extrusion to produce controlled-release tablets. Int. J. Pharm. [Internet]. 2017;519:186-197. Available from: http://dx.doi.org/10.1016/j.ijpharm.2016.12.049.

[36] Jonathan G, Karim A. 3D printing in pharmaceutics: A new tool for designing customized drug delivery systems. Int. J. Pharm. [Internet]. 2016;499:376-394. Available from: http://www.sciencedirect.com/science/article/pii/S0378517315304622.

[37] Vogenberg FR, Isaacson Barash C, Pursel M. Personalized medicine: part 1: evolution and development into theranostics. P T. 2010;35:560-576.

[38] Abubakar AR, Chedi BAZ, Mohammed KG, et al. Drug interaction and its implication in clinical practice and personalized medicine. Natl. J. Physiol. Pharm. Pharmacol. 2015. p. 343-349.

[39] Selmin F, Musazzi UM, Cilurzo F, et al. Alternatives when an authorized medicinal product is not available. Med. Access @ Point Care. 2017;1:16-21.

[40] Eleftheriadis GK, Katsiotis CS, Andreadis DA, et al. Inkjet printing of a thermolabile model drug onto FDM-printed substrates: formulation and evaluation. Drug Dev. Ind. Pharm. [Internet]. 2020;46:1253-1264. Available from: https://doi.org/10.1080/03639045.2020.1788062.

[41] El Aita I, Ponsar H, Quodbach J. A Critical Review on 3D-printed Dosage Forms. Curr. Pharm. Des. 2018;24:4957-4978.

[42] Trenfield S, Basit A, Goyanes A. INNOVATIONS IN 3D: Printed pharmaceuticals. drug Deliv. 2020;2020:45-49.

[43] Goyanes A, Madla CM, Umerji A, et al. Automated therapy preparation of isoleucine formulations using 3D printing for the treatment of MSUD:

First single-centre, prospective, crossover study in patients. Int. J. Pharm. [Internet]. 2019;567:118497. Available from: https://doi.org/10.1016/j.ijpharm.2019.118497.

[44] Fernández-García R, Prada M, Bolás-Fernández F, et al. Oral Fixed-Dose Combination Pharmaceutical Products: Industrial Manufacturing Versus Personalized 3D Printing. Pharm. Res. 2020;37.

[45] Alomari M, Vuddanda PR, Trenfield SJ, et al. Printing T3and T4oral drug combinations as a novel strategy for hypothyroidism. Int. J. Pharm. 2018;549:363-369.

[46] Alhnan MA, Okwuosa TC, Sadia M, et al. Emergence of 3D Printed Dosage Forms: Opportunities and Challenges. Pharm. Res. [Internet]. 2016;33:1817-1832. Available from: http://dx.doi.org/10.1007/s11095-016-1933-1.

[47] FDA Technical Considerations for Additive Manufactured Medical Devices: available at: www.fda.gov/files/medical%20devices/published/Technical-Considerations-for-Additive-Manufactured-Medical-Devices---Guidance-for-Industry-and-Food-and-Drug-Administration-Staff.pdf. Accessed June 28, 2021.

[48] Patel MA, AbouGhaly MHH, Schryer-Praga J V., et al. The effect of ionotropic gelation residence time on alginate cross-linking and properties. Carbohydr. Polym. [Internet]. 2017;155:362-371. Available from: http://dx.doi.org/10.1016/j.carbpol.2016.08.095.

[49] Dalheim M, Vanacker J, Najmi MA, et al. Efficient functionalization of alginate biomaterials. Biomaterials. 2016;80:146-156.

[50] Boateng JS, Stevens HNE, Eccleston GM, et al. Development and mechanical characterization of solvent-cast polymeric films as potential drug delivery systems to mucosal surfaces. Drug Dev. Ind. Pharm. 2009;35: 986-996.

[51] Pacheco MS, Barbieri D, da Silva CF, et al. A review on orally disintegrating films (ODFs) made from natural polymers such as pullulan, maltodextrin, starch, and others. Int. J. Biol. Macromol. [Internet]. 2021;178:504-513. Available from: https://doi.org/10.1016/j.ijbiomac.2021.02.180.

Chapter 8

Curcumin-Alginate Mixed Nanocomposite: An Evolving Therapy for Wound Healing

Sulata Sahu and Bairagi C. Mallick

Abstract

A lot of advancements have taken place in the wound dressing materials and in wound healing process. Alginate based wound dressings materials are more preferable due to their biocompatibility and non-toxic unique biological characteristics. There's always a need to increase the efficacy of alginates by combining with other biopolymers like chitosan, collagen and cellulose etc. However, the recent trend towards the natural and herbal bio-compounds are more likely attracting to develop alginate based wound dressing materials with higher efficiency, antimicrobial and anti-inflammatory potency. Out of many natural compounds tested, curcumin has shown high potency and more effectively used for wound healing purpose. Due to curcumin's bio-medical properties it has been used as a vital ingredient combined with alginate and other biopolymers to prepare wound dressing materials. Based on the available literatures, this review chapter on alginate-curcumin nanocomposite will help the reader to develop better wound healing materials with evolutionary therapeutic applications.

Keywords: Alginate, curcumin, nanocomposite, wound healing dressing materials

1. Introduction

Despite the progress in wound dressing materials and also the level of expertise within the skilled discipline, the wound healing management still remains inconclusive [1]. The reason could be the price management of wounds is more and also the increasing population size and metropolitan lifestyle. These facts indicates towards the increasing demand for development of wound dressings that are effective, acceptable and cheap. In the year 2012, 184 million pound was spent on wound dressing merchandise in England [2]. Similarly, in the USA, annually approximately 20 billion dollar is spent on the management of chronic wounds [3, 4]. There are some barriers like educational, organisational, clinical and psychosocial for successful and effective treatment of wounds [5]. Biomaterials, biopolymers, synthetic polymers are the raw materials for preparation of wound dressing materials. Properties of biopolymers should be non-toxic, easily accessible, perishable, biocompatible and non-immunogenic like chitosan, alginate, fucoidan, mucopolysaccharide etc. [6, 7]. Due to the non-toxicity, biocompatibility, non-immunogenicity, affordability and high absorption capability specific and selective properties of polymers, alginates are best suitable used for dressing materials for wound dressing purpose [8]. However, due to poor mechanical property of biopolymers they are incorporated with artificial polymers to boost their mechanical properties and tailored to change their degradation mechanism [1, 6, 7].

Alginate is a polysaccharide, generally used as bio-polymericmaterial for wound dressing [1]. It is a known and naturally collected biopolymer used in the management of wound dressings because of its selective characters like biocompatibility, nature of gelling and swelling, which creates moist and microenvironment at the damaged site enhancing the healing mechanism and declining the period of heling [1, 9]. The special properties of alginate convert it a model biopolymer with potential value, which might overcome the drawbacks related to other biopolymers used as wound dressing material [9, 10]. Alginate has already been incorporated within selective biomaterials for formulations of hydrogels, films, foams, gels, wafers, nanofibers, topical formulations, and several other novel systems for wound dressing [1, 9, 11]. Alginates are also utilised to prepare sponges as wound dressing products [1, 12]. The haemostatic property of alginate made appropriate for bleeding wounds also [13]. As alginate based hydrogel products are extensively used in tissue recreation and also acceptable in highly damaged wounds, these systems are most preferable in case of wound healing process [12, 14, 15].

2. Alginate

British Chemist E.C.C. Stanford in 1881 first described Alginate and available as the most common and abundant polysaccharide in the brown algae having 40% of the dry matter. It is composed of sodium, calcium, magnesium, strontium and barium ions in a gel form located in the intercellular matrix. It is mostly utilised for industrial purpose because of its capacity of retaining water, gelling, viscosifying and stabilising properties [16].

Alginate has been broadly investigated and employed for many biomedical implementations because of its biocompatibility, low toxicity, low cost, and mild gelation by adding of divalent cations such as Ca^{2+} [17, 18]. It is different from the other hydrocolloids like agar and carrageenan as it is collected from brown seaweed (Phaeophycean) exactly from outer layer cell wall because inner layer mostly made up of cellulose [19]. These alginate molecules allow both flexibility and mechanical strength to the algae [20, 21]. Apart from brown algae, it can also collected from different bacteria species such as *Azotobacter* and *Pseudomonas* [22]. Brown algal species like *Laminaria hyperborean*, *Macrocystis pyrifera*, *Laminaria digitata* and *Ascophyllumnodosum* are extensively used for commercial alginate preparation, while species like *Sargassum spp.*, *Laminaria japonica*, *Ecklonia maxima* and *Lessonia nigrescens*are utilised only if other brown seaweeds not accessible as these gave low and weak production of alginates [23, 24]. Although alginates can be produced both from algae and bacteria, mostly algae are used for commercial production [25].

3. Structure and composition

Alginates are unbranched polysaccharides that are made up of 1, 4 linked β-D-mannuronic acid (M) and its C-5 epimer, α-L-guluronic acid (G). It is composed of M sequences (M-blocks) and G sequences (G-blocks) distributed with MG sequences (MG-blocks) [25]. The D-mannuronate was thought to be the main component of alginate before Fischer and Dörfel discovered the α-l-guluronate residue [26]. Later, fractional precipitation with manganese and calcium salts revealed that alginates are also block copolymers with a different ratio of guluronate to mannuronate depending on the natural source [27]. Alginate is now understood to be a group of linear copolymers made up of 1,4-linked β-D-mannuronate (M) and α-l-guluronate (G) residues. Consecutive G residues (GGGGGG), consecutive M residues (MMMMMM) and alternating M and G residues (GMGMGM) make up the blocks (**Figure 1**). The M and G contents of alginates extracted from various sources vary, as does the length of each block, and over 200 different alginates are currently manufactured [28]. The G-block content of L-hyperborean stems is 60%, compared to 14 to 31% for other commercially available alginates [29]. Only the G-blocks of alginate are thought to engage in the formation of hydrogels by intermolecular cross-linking with divalent cations (e.g., Ca^{2+}). The composition (i.e., M/G ratio), sequence, G-block length, and molecular weight of alginate and its resulting hydrogels are thus critical factors influencing their physical properties [30]. The length of the G-block and the molecular weight of alginate gels are usually increased to improve mechanical properties. Different alginate sources manufacture polymers with a variety of chemical structures, for example, bacterial alginate derived from Azotobacter has a high concentration of G-blocks and its gels are relatively stiff [31]. The stability of the gels, the rate of drug release from gels, and the phenotype and function of cells encapsulated in alginate gels are all influenced by their physical properties [18].

Commercially available sodium alginates have molecular weights ranging from 32,000 to 400,000 g/mol. For sodium alginate in 0.1 M NaCl solution at 25°C, the parameters of the Mark–Houwink relationship; [η] = KMv a) are K = 2 103 and a = 0.97, where [η] is intrinsic viscosity (mL/g) and Mv is viscosity-average molecular weight (g/mol) [32]. As the carboxylate groups in the alginate backbone become protonated and form hydrogen bonds, the viscosity of alginate solutions increases as the pH decreases, peaking around pH 3 to 3.5. The physical properties of the resulting gels can be improved by increasing the molecular weight of alginate. An alginate solution made from a high molecular weight polymer, on the other hand, becomes extremely viscous, making it difficult to work with [33]. The pre-gel solution viscosity and post-gelling stiffness can be regulated independently by manipulating the molecular weight and its distribution. By combining high and low molecular weight alginate polymers, the elastic modulus of gels can be greatly increased while the viscosity of the solution is minimally increased [18, 34].

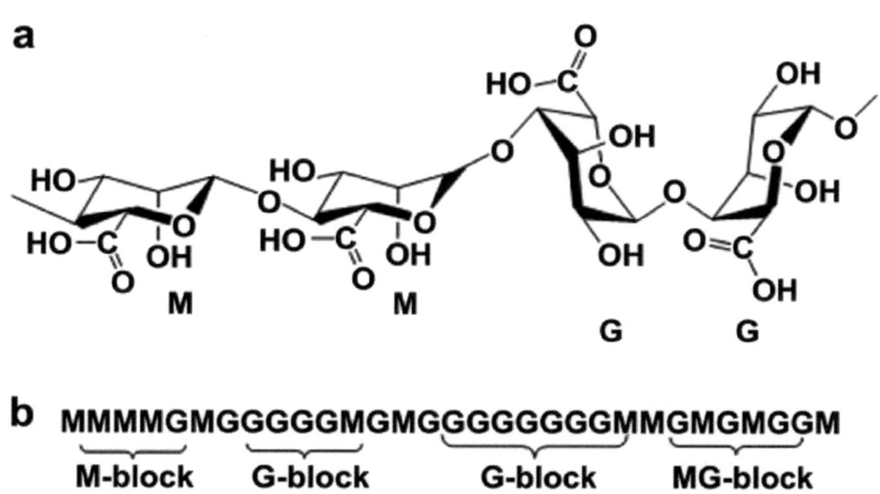

Figure 1.
Diagrammatic representation of alginates: (a) molecular chain, (b) M and G block distribution [25].

4. Biosynthesis

Several excellent research articles [22, 31] have analysed recent progress in bacterial alginates biosynthesis. The oxidation of a carbon source to acetyl-CoA, which enters the TCA cycle and is converted to fructose-6-phosphate through gluconeogenesis, is the first step in alginate biosynthesis. Fructose-6-phosphate is then transformed to GDP-mannuronic acid, which is a precursor to alginate synthesis, through a sequence of biosynthetic transformations. The biosynthetic process can be divided into four stages in general: (a) synthesis of the GDP-mannuronic acid precursor; (b) cytoplasmic membrane transfer and polymerisation of polymannuronic acid; (c) periplasmic transfer and modification; and (d) export through the outer membrane. Polymannuronic acid is acetylated at the O^{-2} and/or O^{-3} positions by multiple transacetylases at stage (c), resulting in post-polymerisation modification of alginates [35, 36]. A family of epimerase enzymes then performs epimerization to transform certain non-acetylated M residues to G residues [37–40]. Finally, transmembrane porin sallow alginate to exit the cell [25].

Alginate is generally available in sodium, potassium, or calcium forms in the market. Both alginate salts are widely used as biopolymers in medical, biomedical, medicinal, and cosmeceutical applications [1, 41–43]. In the presence of divalent and trivalent metal cations in the aqueous setting, sodium alginate can form ionotropic hydro-gelled matrices [44, 45]. Alginate is a biopolymeric excipient widely used in pharmaceutical and biomedical products such as capsules, hydrogels, gels, managed release systems, beads, bio-adhesives, pellets, patches, microparticles, and nanoparticles as an emulsifying agent, disintegrant, thickener, coating content, stabiliser, and so on [46–49]. Based on the cross-linkers used and cross-linking techniques used, alginates can entrap drug molecules and release them in a rate-controlled pattern. It's used in managed drug release systems and targeted drug delivery systems to ensure that drugs have the best bioavailability at their target sites [12]. Sodium alginate, calcium alginate, and alginate derivatives have also been used in the manufacture of wound dressings. Water absorption power, swelling and gelling capabilities, ability to be crosslinked, controllable porosity, biodegradability, and biocompatibility nonimmunogenic, haemostatic nature, bioactivity (support the proliferation process), bio-similarity to extracellular matrices, bio-adhesivity, ability to encapsulate drugs and control of drug releasing, cost-effective, etc. are all important properties that make these biopolymeric materials ideal for wound dressing formulations. Alginate has the ability to absorb up to 20 times its weight in body fluids and liquids, creating a hydrophilic gel. Alginate's excellent gelling properties made it suitable for wound dressing applications [12, 50].

5. Alginate nanoparticles

Nanoaggregates, nano-capsules, and nanospheres are nanoscale systems (10 1000 nm) that can contain enzymes, medicines, and other compounds by dissolving, encasing, or adding them to the matrix of the particle. Different methods used for prepare nanoparticles involves: (a) nanoaggregates, which are nanoscale colloidal structures in which the drug is mechanically distributed and can take on a variety of morphologies; (b) nano-capsules, which are vesicular systems in which the drug is confined to an oily or aqueous liquid centre enclosed by a polymeric membrane, (c) nanospheres, which are spherical particles with a gelled interior in which the entrapped component is physically distributed, or (d) nano-capsules with a structured interior, which are a hybrid of a nano-capsule and a nanosphere; they are made by first preparing a nanosphere and then forming an additional shell on the nanosphere's interface [51–55]. As previously mentioned, there are two methods for producing alginate nanoparticles: (i) complexation, which occurs in an aqueous solution to form alginate nano-aggregates or on the interface of an oil droplet to form alginate nano-capsules; a crosslinker (e.g., calcium from calcium chloride) is used for complexation of alginate, and complexation may also occur by mixing alginate with an oppositely charged polyelectrolyte (*e.g.*, poly-L-lysine); and (ii) alginate in-oil emulsification, followed by external or internal alginate emulsion droplet gelation, resulting in alginate nanospheres [56]. We can summarise the methods used in synthesising alginate-based nanomaterials as-controlled jellification using Ca^{2+} ions, formation of polyionic complexes through ionotropic gelation via intermolecular interactions, spray drying followed by crosslinking, alginate nanoaggregates through self-assembly, fabrication of nanocomposite fibrous scaffolds and nanoparticles respectively by electrospinning and electro-spraying, fabrication of nanocomposite fibrous scaffolds by thermally-induced phase separation and formation of alginate nanoparticle using a microfluidics- aided polyelectrolyte Complexation [57].

6. Wound

By functioning as a physical/chemical barrier [9], skin the largest organ of body protects the internal organs and is responsible for protecting against infections and dehydration from environmental aggressions [58]. However, it is the human body's most often damaged organ. Skin (acuteand chronic) wound defence from pathogens when it is injured (diabetes mellitus, chronic venous, arterial insufficiency, immunological and other infections) is a very difficult problem for the recovery of the injured skin. Wound is described by the Wound Healing Society as the disturbance of normal anatomical structure and functions [59]. The prevalence and severity of wounds prompted researchers to focus on wound healing management studies, as well as the demand for wound dressings. Accidents, burns, surgical operations, and violent impact all cause breaks or lacerations in the skin's membrane layers, causing damage to underlying tissues or disruption of cellular integrity, resulting in wounds [60]. Physical, mechanical, thermal, biochemical, surgical, or metabolism-related problems can all result in wounds [61–64]. If left untreated, the wound can develop sepsis, necessitating amputation or even death [60].

Wounds are graded based on (a) the time and nature of the wound healing process, (b) the depth of the wound injury, and (c) the appearance of the wounds [12]. Acute and chronic wounds are the two types of wounds that are often encountered. An acute wound is a skin injury that happens unexpectedly rather than over time, and heals in 1–12 weeks, depending on the type of the wound. It should be noted that for optimal health and cost, the healing of an acute wound (which may occur in a variety of ways) is preferable for recovery [65]. Chronic wounds can be caused by a variety of factors, including venous insufficiency, arterial perfusion, diabetes, and so on. Chronic wounds do not heal in the expected time frame because they are more vulnerable to infections and are more difficult to treat [9, 66].

Chronic wounds are divided into four categories based on their aetiology: (a) decubitus ulcers (bedsores), (b) diabetic ulcers, (c) venous ulcers (leg ulcers), and (d) arterial insufficiency ulcers [67]. Wounds are also divided as (i) superficial wounds (injury of surface of epidermis only), (ii) partial-thickness wounds (injury on both epidermis and deep into dermis like blood vessels, sweat ducts etc.), and (iii) full-thickness wounds (injury on underlying the subcutaneous fatty layers along with the layers of epidermis and dermis), (iv) on the other hand, wounds may be categorised as necrotic, sloughy, granulating, epithelializing, contaminated,

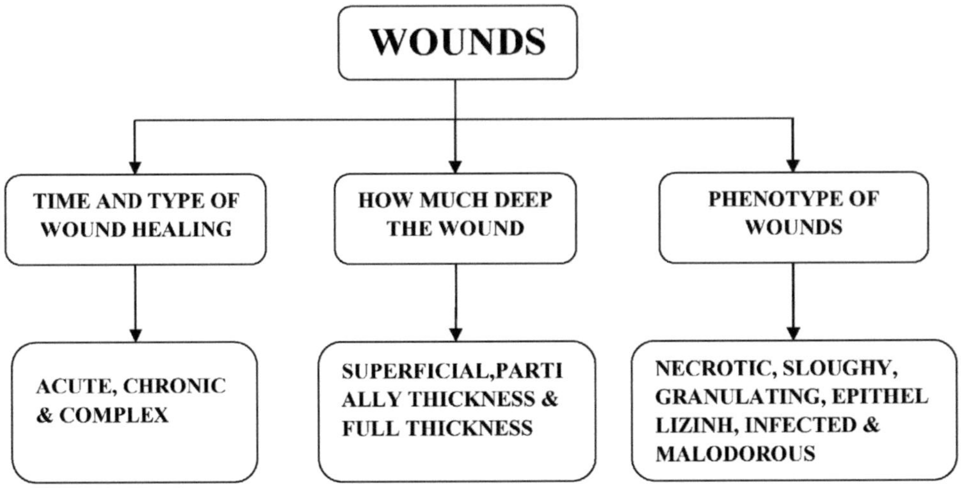

Figure 2.
Classification of wounds [12].

or malodorous depending on their appearance (**Figure 2**). Complex wounds, which are either acute or chronic wounds that are difficult to heal are another new type of wounds [68–70]. Burn injuries are a form of serious wound that is frequently painful and physically damaging, accompanied by pain and inflammation. Burn wounds may often result in prolonged sensory difficulties, serious illness, and mortality as a result of long hospitalisation and recovery [71, 72]. Medical Treatment forburn wounds is still difficult, and several studies are being conducted to develop better therapeutic aids. For successful care, current industrial wound dressing products and engineered skin replacements are being pushed towards a functionalized wound dressing strategy [72].

7. Wound dressing classification

Traditional, biomaterial-based, interactive, and bioactive dressings are the four types of wound dressings [73]. Traditional dressings, also called as passive wound dressings, are utilised to shield wounds from the environmental contact while also preventing bleeding [73, 74]. Traditional dressings include gauze and gauze-cotton composites, both of which have a high absorptive quality. These can, however, cause bleeding, having poor vapour permeability, and damage the freshly formed epithelium when removed. Exudates leaking from these dressings may also cause bacterial infections. Allografts, tissue derivatives, and xenografts are three types of biomaterial-based wound dressings [73]. Allografts can be defined as skin fragments collected from donors that may be fresh or freeze-dried, and their usages was restricted due to an immune response that causes the body to reject them. There's also the possibility of infection and disease transmission [75, 76]. They are costly and have a short shelf life. Tissue derivatives are made from collagen, but their use is restricted due to the possibility of infection over a long time. Artificial dressings, also called immersive wound dressings, are made up of gels, foams, films, sprays, composites, and other materials [73, 74]. Biopolymers and synthetic polymers are used to prepare them. Alginate, chitosan, gelatine, and other biopolymers are commonly used. Wound dressings may also be categorised as bioactive wound dressings, with alginates, collagens, hydro-fibres, and hydrocolloids being examples. Wound healing is aided by the addition of growth factors and antimicrobials agents.

A lot of biopolymers are now being frequently used to make the wound healing materials for different type of wounds [1, 73].

8. Wound healing process

Haemostasis, inflammation, migration, proliferation, and remodelling are the five stages of wound healing (**Figure 3**), [3, 68–80]. When an injury to the skin occurs, haemostasis and inflammation occur. Fibrinogen, a major component of the skin's connective tissue, aids in the coagulation of exudates and blood clotting in wounds to avoid bleeding [3, 73, 80, 81]. The inflammatory process occurs concurrently with the haemostasis phase, in which the phagocytic cells release proteases and reactive oxygen species, which clean the wound of debris and protect it from bacterial infection [3]. Blood monocytes transform into tissue macrophages at the wound site, releasing growth factors and cytokines that attract fibroblasts, endothelial cells, and keratinocytes to help repair damaged blood vessels. The epithelial cells migrate towards the wound site to replace dead cells during the migration process. The wound is fully covered by epithelium during the proliferation stage, and formation of granulating tissues starts. Tissue remodelling is the final step, in which fibroblasts fully cover the wound's surface as a new layer of skin. There is no evidence of wound in this process, which is also known as the maturation phase [1, 3, 78].

Damaged skin is vulnerable to microbial infection and is unable to protect the physiological functions of internal organs [82–84]. Wound dressing will hasten skin's physiological regeneration while also preventing infections and dehydration at the wound surface [85]. Furthermore, an ideal wound dressing should have good biocompatibility [86]. As a result, a variety of wound dressings in various types have recently appeared, the majority of which are made up of natural macromolecules such as chitosan, cellulose, alginate, and collagen [87–89]. Alginate is widely used in the pharmaceutical and biomedical fields as a natural biocompatible polysaccharide. Alginate has also been used as a wound dressing in the past. While alginate dressings can help with wound healing, they have poor haemostatic properties, particularly when it comes to massive haemorrhage [90]. Collagen which is the most abundant protein in animals, and it can be found in almost all soft and hard connective tissues. Collagen is also a vital constituent of the extracellular matrix (ECM), which organises itself around the cells in a logical manner.

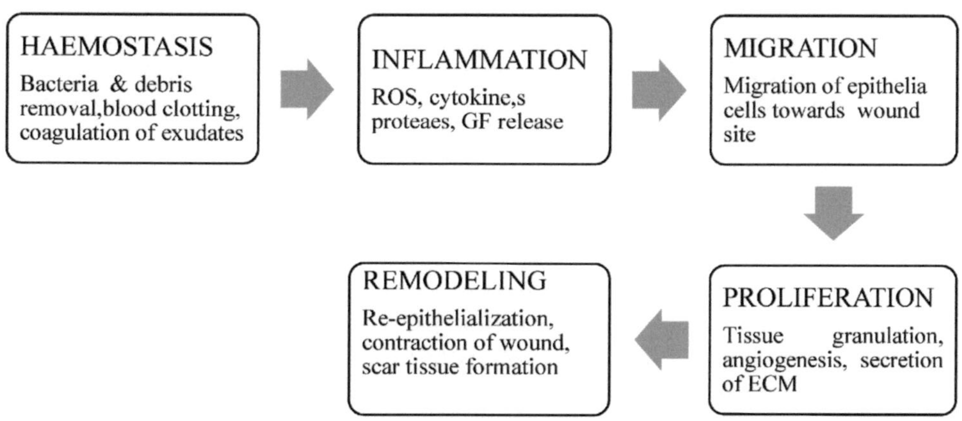

Figure 3.
Steps in wound healing processes [1].

Collagen has a high biocompatibility, a strong haemostatic activity, a higher water absorbing potential, and a low immunogenicity [91]. Collagen can also facilitate cell attachment and proliferation, affecting cellular activity and tissue function further. Collagen-based wound dressings were found to have a number of advantages in studies. The addition of collagen to alginate significantly improved the hydrogel's properties [92]. Meanwhile, the collagen–alginate composite had considerable water absorption and mechanical properties, allowing it to keep the wound surface moist as well as maintain local tissue and encourage cell adhesion and growth [93, 94]. On the other hand Collagen and alginate, on the other hand, display no antimicrobial activity. As a result, antimicrobial functionalization of collagen–alginate dressings is critical to promote wound healing and prevent secondary infection [95]. The abundance of cellulose in the biosphere, especially in herbal and bacterial sources, has made this natural polysaccharide more available. To make cellulose, D-glucose units are connected together by glycoside bonds of $\beta(1 \rightarrow 4)$to form the formula $(C_6H_{10}O_5)_n$, which has three hydroxyl groups for each unite. These functional groups are specifically targeted for cellulose modifications in order to enhance biomedical applications such as wound dressing manufacturing. For example, in the case of modified bacterial cellulose/keratin nanofibrous mats through a hydrogel of tragacanth natural gum, increased fibroblast cell attachment and proliferation were observed [96]. Due to its excellent biocompatibility, nontoxicity, antibacterial properties, and haemostasis, chitosan is also considered a highly favourable material, while gelatine has excellent film-forming and water absorbing properties [86, 97–99]. In addition to all as a natural and herbal medicine, curcumin was mixed into the chitosan and alginate sponge (CA sponge) as a multifunctional agent to prevent wound infection. Curcumin (CUR), a polyphenol derived from turmeric, is a well-known wound-healing agent that has been shown to have antimicrobial, antioxidant, and anti-inflammatory properties [100–102].

9. Curcumin

Curcumin is one of the most important curcuminoids found in *Curcuma longa*. Turmeric is the common name for it. Curcumin, also known as diferuloylmethane or 1,7-bis (4-hydroxy-3-methoxyphenyl)-1,6-heptadiene-3,5-dione, is a polyphenolic phytoconstituent with a low molecular weight that occurs naturally [103, 104]. It is a member of the Zingiberaceae ginger family (chemically named diferuloyl methane). Turmeric has been used by the South Asians for treating diseases and skin problems since ancient times. In reality, it is a widely used spice and a staple in their diet [105, 106]. Curcumin is a powerful anti-inflammatory agent. It is also said to inhibit tumour growth to some degree. As a result, the scientific community has taken a keen interest in this material, and a significant number of experiments have been carried out with it for a variety of purposes. This is aided by the fact that curcumin is nontoxic when taken at a dose of 4 g/day for up to 120 days [107–109] (**Figure 4**).

Curcumin's pharmacological properties, including anti-inflammatory [111], antimicrobial [112], antiviral [113], anticancer [114], antioxidant [115], chemosensitizer [116], radiosensitizer [35], and wound healing activities [117, 118], indicate that it has potential in preclinical cell culture and animal studies. Topical usage of curcumin have been shown to enhance healing of wounds and protect tissues from oxidative damage by acting as an anti-inflammatory antioxidant (free radical scavenging activity), inducing detoxification enzymes, and providing defence against degenerative disease in patients [119, 120].

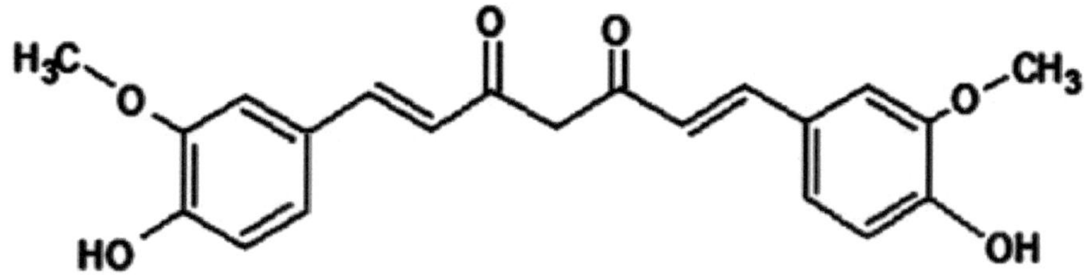

Figure 4.
Structure of curcumin (diferuloylmethane) [110].

10. Pharmacological activity of curcumin

Curcumin has been widely regarded as a "wonder drug of the future" due to its exceptional ability in preventing and treating many incurable illnesses, according to several clinical and preclinical studies performed on it over the past decades [121]. Curcumin has been studied for its anticancer properties for the past 50 years. According to some reports, Curcumin was shown to have chemotherapeutic impact on breast cancer [122], colon cancer [123], stomach cancer [124], liver cancer [125], and oral cancer [126]. It's effects on diseases including diabetes [127], Alzheimer [128], gastric ulcer [129], rheumatoid arthritis [130], psoriasis [131], and HIV [132] were studied in other studies, revealing its potential as a therapeutic agent. Curcumin's effects on the cardiovascular system, as well as pulmonary and metabolic disorders, have been studied further [133]. Curcumin's antimicrobial [134], antiviral [135], antioxidant [136], anti-inflammatory [137], and wound healing properties have also been widely recorded [117, 138].

Despite all of curcumin's potential usage in medicine, pharmacological studies have shown a number of disadvantages. Curcumin was discovered to have low absorption, poor solubility, and a low distribution property. As a result, curcumin is poorly absorbed in the body and cannot be used to treat diseases [139, 140]. It has a low bioavailability due to its rapid metabolism, elimination, poor tissue distribution, and low serum concentration [140, 141]. According to studies, curcumin bioavailability varies depending on the route of administration. In *in-vivo* study on mouse sample, it was discovered that when curcumin was administered intravenously, it was absorbed the most in the serum, while when curcumin was given orally, it was absorbed the least [142]. Researchers have been trying to increase the bioavailability of curcumin for the past three decades. They have devised a number of methods to improve the medicinal use of curcumin. Curcumin has been studied in various formulations to see whether they can solve the problem [143]. We have concentrated in this chapter on nanotechnology-based techniques that can aid in the delivery of curcumin and thus alleviate its disadvantages [110].

11. Curcumin as a nanoparticle

Nanoparticles (NPs) are now being researched and used in a variety of medical fields to allow for targeted drug delivery, lower active agent dosing, combination therapy, reduced side effects, and the use of more potent drugs that cannot be clinically used by traditional drug delivery. Because of its many unique characteristics, nanotechnology has become a major area of interest in response to the growing threat of microbial drug resistance. The ability to overcome barriers and gain access to biological molecules and more specifically, microorganisms is enabled by the physical and chemical properties of NPs including their high surface-to-volume ratio and small size [144]. The mechanisms other than antibiotic arsenal, which has established resistance, this direct association with microbial cell membranes/walls and main proteins/enzymes can both inhibit pathogen growth and/or instigate cell death. Furthermore, the scale, shape, and chemical properties of NPs can be altered to promote these molecular interactions, thereby maximising their ultimate function [144, 145].

Curcumin encapsulation in a nanoparticle platform is a feasible and beneficial method of delivering it. Nanoparticles can pass through the skin barrier and intracellularly due to their small size and high surface-to-volume ratio [146], making them suitable for topical drug delivery. Since the potential maximum volume of substance is never in contact with skin at one time, toxicity is limited by the slow, continuous release of encapsulated contents. Nanoparticles allow the delivery of substances like curcumin that have physicochemical properties that prevent or delay their use in non-encapsulated form. Curcumin nano-formulations have been established for preclinical studies on cancer, inflammation, wound healing, and other conditions, with nano-versus non-encapsulated curcumin demonstrating improved therapeutic efficacy [146, 147]. However, there is minimal evidence in the infection landscape [148], with no *in-vivo* studies reported to date [149].

Nanomaterials have the ability to enhance and change the pharmacodynamic properties of drug molecules with solubility issues, such as curcumin. Curcumin's stability, solubility, and bioavailability can all be improved with nanoparticle-based drug delivery. Physical entrapment, adsorption, and chemical conjugation methods are used to bind drugs to the surface of nanoparticles [140, 143, 150]. Various forms of nanomaterials, such as liposomes, proteins, nano-emulsions, micelles, strong lipid nanoparticles, and polymeric nanoparticles, have been found to be promising carriers for the delivery of curcumin in previous studies. Polymeric nanoparticles are thought to be the most important of all nanoparticle-based delivery methods and are currently being studied for the curcumin delivery [110, 151–153].

12. Alginate-curcumin based systems for wound healing

Acute and chronic wound care is a pressing need, and alginate-based wound dressings provides many benefits over the conventional wound dressings. Alginate-based wound dressings come in a variety of shapes and sizes, including hydrogels, films, foams, nanofibers, membranes, and sponges. Alginate based dressings can absorb wound fluid, resulting in gels that keep the wound site physiologically wet and prevent bacterial infections. Efficacy of these wound dressings has been documented by several researchers [1]. To increase the efficacy of alginate dressing, curcumin was used as a potential antibacterial and ani-inflammatory biomaterial for rapid healing of wounds. In combination with alginate and other biopolymers curcumin increases the potency of wound healing of the wound dressing systems. Many alginate-curcumin systems have been used as drug delivery systems, wound dressing materials, tissue regeneration materials, and other applications in recent years. Curcumin's use in wound healing applications as wound dressing products has been well researched over the last few decades and published in numerous publications [154]. Other than alginate, chitosan was the mostly used biopolymers for preparation of wound care dressing materials. The focused on researches and experiments done with alginate and curcumin and related wound healing dressing systems and summarised in **Table 1**.

There are many attempts to improve the efficiency of wound healing through slow delivery of antioxidants including curcumin from collagen, which also serves as a supporting matrix for the regenerative tissue [167]. In the curcumin

Alginate-curcumin formulations	Composition	Investigation level/stage	Result	Ref.
Sponge	Chitosan, sodium alginate (CA) and curcumin	*in-vivo*	CA-Curcumin sponge can promote the formation of the collagen, thus accelerate the healing process.	[120]
Foam	Sodium alginate and curcumin	*in-vitro*	It can be removed easily from the wound site with wound healing effect.	[155]
Hydrogel	N,O carboxymethyl chitosan/oxidised alginate and nano-curcumin	*in-vivo*	It enhances the re-epithelialization, collagen deposition and accelerate the process of wound healing.	[156]
Polymeric bandage	Chitosan, Sodium alginate, oleic acid and Curcumin	*in-vivo*	It enhances wound healing.	[157]
Microfibre	sodium alginate, gelatine and Curcumin	*in-vivo*	The composite has the potential to wound healing platform.	[158]
Film	Bacterial cellulose, alginate, gelatine and curcumin	*in-vitro*	It has potency for wound care, periodontitis and oral cancer treatment.	[159]
Fabric scaffold	Curcumin and Gymnemasylvestre, graphene oxide, polyhydroxy butyrate-sodium alginate	*in-vivo*	Effective against diabetic wound as well as normal wound.	[160]
Film	Human Elastin-Like Polypeptide (HELP), alginate and curcumin	*in-vitro*	Effective for drug delivery, wound healing, and tissue regeneration.	[161]
Hydrogel	Curcumin, β-cyclodextrin Sodium alginate, Chitosan	*in-vitro*	It inhibits the Gram-negative (*E. coli*) and Gram-positive bacteria (*S. aureus*) growth to heal the wound.	[162]
Sponge	Curcumin, chitosan–alginate and β-cyclodextrin	*in-vivo*	It facilitate cutaneous wound healing.	[163]
Membrane	Alginate membrane, polycaprolactone (PCL) nanoparticles, Curcumin	*in-vivo*	It can accelerate wound healing without removal of the membrane.	[164]
Patch	Sodium alginate/PVA-Titanium dioxide (TiO_2) and curcumin	*in-vitro*	It shows antibacterial and antifungal activities and can be used for wound healing.	[165]
Curcumin Encapsulated particle	Pluronic F127 (PF127), calcium chloride ($CaCl_2$), sodium alginate, chitosan, curcumin	*in-vitro*	It has potency of good drug delivery and further research needed to test the efficacy for wound healing.	[166]

Table 1.
List of experiments on alginate-curcumin nanocomposite wound healing systems.

incorporated collagen matrix (CICM) group, biochemical parameters and histological analysis revealed increased wound healing, increased cell proliferation and effective free radical scavenging. As compared to standard collagen films, CICM films have a higher shrinkage temperature, implying greater hydrothermal stability. Curcumin was found to be bound to collagen without altering its triple helicity in spectroscopic tests. Curcumin absorbs a lot of light in the wavelength range of 300 to 500 nm. In methanol solution, it has a maximum absorption wavelength of 420 nm, which changes to 430 nm in dimethylformamide. As collagen solution is applied to curcumin, however, the maximum wavelength changes from 420to 429 nm. The change in the absorption shift stabilised and the amplitude of the peak increased with rising curcumin concentration, implying that curcumin is affected by the hydrophobic atmosphere. Further, lipid peroxidation method was used to determine the antioxidant performance of CICM [168]. The CICM's *in-vitro* antioxidant activity was tested by using 2,2-azobisisobutyronitrile antioxidant assay. According to antioxidant research, CICM is more effective at quenching free radicals. This concept supports the topical application of CICM as a viable and effective method for supporting dermal wound healing [167].

A nanocomposite sponge was made out of curcumin-chitosan-gelatine at various ratios of chitosan and gelatine [169], and the chemical structure and morphology were characterised using ATR- FTIR and FE-SEM. The water absorption ability, antibacterial activity, *in-vitro* drug release and *in-vivo* wound healing studies were also performed on these sponges using a rabbit excision wound healing model. The combine form of curcumin, chitosan, and gelatine has shown increase water absorption capacity, antibacterial function and wound closure potency. The cytotoxicity of curcumin sponge tested using the indirect MTT method on mouse fibroblast cells (L929) [170] indicates the controlled release of curcumin increased as the gelatine content of the prepared sponge increased. The drug in the sponge with a higher gelatine content is favoured to be released, instead of being dissolved by the sponge according to the partition ratio between sponge phase and water phase. The findings shows cell viability after 24 hours of incubation period with sponge-released medium. The results show that in the presence of medium containing sponge without curcumin and curcumin sponge, cell viability drops to around 90%. In the *in-vivo* analysis, wounds treated with curcumin-composite sponge closed faster than wounds treated with composite sponge without curcumin. According to the results of curcumin drug release, composite sponges with a higher gelatine content had a faster release activity up to 240 minutes. These composite sponges were also discovered to increase wound healing activity by enhancing collagen production *in-vivo*. These obtained results showed that combination of curcumin, chitosan and gelatine could improve the wound healing activity in comparison to chitosan, and gelatine without curcumin [169].

Hydrogels are most preferable system as their biodegradable and controlled drug delivery systems composed of curcumin loaded micelles are widely applied for cutaneous wound repairing [103]. Curcumin with strong antioxidant and anti-inflammatory properties and having high hydrophobicity was encapsulated in polymeric micelles (CurM) for high drug loading and encapsulation efficiency. To improve cutaneous wound healing, Cur-M loaded thermosensitive hydrogel (Cur-M-H) was prepared and applied as a wound dressing materials. At room temperature, Cur-M-H was a free-flowing solution that transformed into a non-flowing gel at body temperature. Cur-M-H system was found to have good tissue adhesiveness and the ability to release curcumin for a long time in vitro studies. It's *in-vivo* wound healing operation was also assessed using linear incision and full-thickness excision wound models. CureMeH-treated mice had a thicker epidermis and higher tensile strength in an incision model. In an excision model, the CureMeH treated group showed improved wound closure. This group also had higher collagen content, stronger granulation, higher wound maturity, a drastic decrease in superoxide dismutase, and a small rise in catalase in both models. CureMeH may also improve cutaneous wound healing, according to histopathologic analysis. Thus the combination of curcumin bioactivity and thermosensitive hydrogel in the in-situ gel-forming composite, facilitates tissue reconstruction processes, implying that the CureMeH composite could be used as a wound dressing for cutaneous wound healing [103] .

According to past reports, materials commonly used to treat burn-wound infections are limited by their incomplete microbial coverage, toxicity, insufficient penetration and more resistance to antibiotics. Curcumin is a naturally occurring compound that has antimicrobial, anti-inflammatory and wound-healing properties. Working through a variety of pathways, is less likely than current antibiotics to select for resistant bacteria. But curcumin's weak aqueous solubility and rapid degradation profile make it difficult to use; however, nanoparticle encapsulation solves this problem and allows curcumin to be delivered to the skin for longer periods of time [149]. So, curcumin nanoparticles (Curc-NP) prepared by the modified sol–gel methods [171] is a major step forward in the treatment of contaminated burn wounds, as it reduces bacterial load while also improving wound healing. Curcumin nanoparticles (Curc-NP) inhibited methicillin-resistant *Staphylococcus aureus* (MRSA) and *Pseudomonas aeruginosa* growth *in-vitro* in a dose-dependent manner, and inhibited MRSA growth and improved wound healing in an *in-vivo* murine wound model. The sensitivity of murine PAM212 keratinocytes to curcumin determined using a semiquantitative FDA (fluorescein diacetate) metabolic assay showed no toxicity to the side tissues. These outcomes indicates, Curc-NPs may rise as a novel topical antimicrobial agent over the conventional antibiotics and wound healing adjuvant for the treatment of infected burn-wounds and other cutaneous injuries [149].

A novel collagen scaffold called nanohybrid-scaffold prepared by incorporating curcumin (CUR) in chitosan nanoparticles (CSNPs) to improve the stability and solubility of the scaffold followed by permeation of prepared CUR-CSNPs have shown better tissue regeneration application [172]. This novel nanohybrid collagen scaffold tested for morphology changes, biodegradability, biocompatibility, *in-vitro* drug release and *in-vivo* wound healing studies have shown high potency. *In-vitro*, the nanohybrid scaffold performed well in terms of water absorption, biocompatibility, and long-term drug availability. Whereas, *in in-vivo* wound closure analysis showed that wounds treated with nanohybrid scaffolds contracted substantially faster than wounds treated with control and placebo scaffolds. Thus the study indicate that in the nanohybrid scaffold community, full epithelialization with thick granulation tissue formation occurs, while there was no compact collagen deposition in the placebo scaffold group and the presence of inflammatory cells in the control group. These suggest that, combining CUR (anti-inflammatory and antioxidant), chitosan (sustain drug carrier, wound healing), and collagen (established wound healer as scaffold) is a promising strategy for addressing various pathological manifestations of diabetic wounds and improving wound healing capability [172].

Considering alginate as a vital ingredient and an unavoidable wound dressing system, a lot of steps have been taken place to develop a proper system having high efficiency for wound healing and low toxicity material. Hence, an ethereal, pliable and biodegradable sponge, composed of chitosan (CS) and sodium alginate (SA) in different ratio and curcumin was prepared and its sponginess, chemical composition and morphology characterised using ATR-FTIR and FE-SEM are similar as reported for other materials for wound dressing purpose [120]. The sponges of all kinds had biodegradable quality and the degree of crosslinking influences the release of curcumin from the sponges. The sponges' water absorption capacity ranged from 10 to 43%. The wound healing test performed on Sprague–Dawley (SD) rats shown that the CA sponges have a continuous release activity for up to 20 days based on the effects of drug release of curcumin. This conclude that the CA sponge acts as an effective drug support for long-term release of curcumin up to 20 days at a time. The collagen bundles in the CA-curcumin sponge-treated wound were thicker than those in the gauze and CA sponge-treated wounds, and they were compact and well-aligned. The developed material, curcumin sponge can promote collagen production, which speeds up the healing process and has a positive impact on wound healing process [120].

Curcumin loaded to alginate foams with low cross-linking was developed based on the previously patented method for the treatment of infected wounds [155, 173]. Curcumin-loaded foam demonstrated a longer hydration time and the amount of curcumin released was adequate, for curcumin-mediated phototoxicity of viable *E. faecalis* cells *in-vitro*. *E. coli*, on the other hand, was less vulnerable to photokilling when curcumin-loaded foams were used, and this was affected by the curcumin solubilizers used in the foams. Only foams containing PEG-400 as the curcumin solubilizer with visible light radiation (29 J/cm^2) caused 81% inhibition of the viable *E. coli* cells to be inactivated. When the foams were exposed to the physiological solution, they quickly hydrated and stayed intact after the loaded curcumin was released, implying that they can be withdrawn from the wound site without causing tissue harm prior to irradiation, reducing light attenuation in photodynamic therapy (PDT) [155].

An injectable chitosan and alginate derivates based on hydrogel incorporating nano-curcumin was also prepared in view of a potential wound dressing with enhanced healing efficacy [156]. Nanocurcumin with a particle size of about 40 nm prepared by using methoxy poly(ethylene glycol)-b-poly(−caprolactone) copolymer (MPEG-PCL) as a carrier, followed by a simple nano-precipitation process and integrated into N,O carboxymethyl chitosan/oxidised alginate (CCS-OA) hydrogel shown slowly and continuously release of curcumin from the CCS-OA hydrogel to promote fibroblast proliferation, capillary development, and collagen output, both of which can have a major impact on the healing processes. The encapsulated nano-curcumin shown slowl released from the CCS-OA hydrogel in an *in-vitro* release analysis (with diffusion controllable release at first, followed by the corrosion manner) of hydrogel at terminal phase. According to the histopathology repots, the hydrogels injected into rat dorsal wounds as part of an *in-vivo* wound healing trial has shown significant improve in epidermis re-epithelialization generation and collagen deposition in wound tissue. The measured DNA, protein, and hydroxyproline content of wound tissue indicate that the nano-curcumin and CCS-OA hydrogel together significantly speed up the wound healing process. It was hypothesised that the nano-curcumin/CCS-OA hydrogel have a wide range of applications in wound healing [156].

On the other side a complete investigation of curcumin loaded oleic acid based polymeric bandage and its therapeutic potential in dermal wound healing in a rat model was carried out [157] to show that curcumin has high wound healing potency in various animal models, and a curcumin-loaded oleic acid-based polymeric (COP) bandage was developed to increase curcumin efficacy in the healing region. Due to its effective free radical scavenging properties, biochemical parameters and histological examination showed increased wound reduction and enhanced cell proliferation in COP bandage treated groups. Researchers believe this is due to increased intercellular curcumin retention and, as a result, an improved anti-inflammatory activity by quenching reactive oxygen species (ROS). Early implementation of fibroblasts and differentiation (increased amount of smooth muscle actin) resulted in a comparative acceleration of wound healing. The COP bandage can effectively quench free radicals, resulting in decreased antioxidative enzyme activity. The mechanism is potent enough to reduce the inflammatory response mediated by the NFB pathway during wound healing, as evidenced from the mRNA and protein levels analysis. In light of this, it's possible that the curcumin-loaded polymeric bandage may have a new therapeutic use in clinical settings for cutaneous wound healing [157].

There are also studies which stated about characterisation and optimisation of different biodegradable and biocompatible formulations of curcumin encapsulated particles, in order to enhance the efficiency of curcumin wound healing effect [166]. The optimised curcumin particles ranged in size from 1286 nm to 1485 nm, with a 75% encapsulation performance. With a Polydispersity Index (PDI) of 0.4, the zeta potential showed values ranging from −7.20 to −7.96. The efficient fabrication and encapsulation of curcumin in the polymeric matrix, which had been fabricated in rod form, was ensured by physical characterisation using HR-TEM imaging. For curcumin particles, the release profile was biphasic, with an initial burst accompanied by a steady release pattern. *In-vitro* cytotoxicity assays and microscopic imaging verified the safety of the curcumin particle concentration used, which was less than 25 g/ml. Furthermore, the findings of a cellular uptake analysis confirmed that curcumin particles were internalised. Overall, the existing biocompatible and biodegradable curcumin encapsulation formulations have the potential to be used as a drug delivery vehicle for curcumin, according to this thesis. More evidence of this curcumin encapsulated particle's ability to improve wound healing is also required [166].

Incorporation of nanoforms of metal oxides into the wound care materials improves their antibacterial efficacy and thus used in wound healing process. This has open up a new window towards the use of metal oxide nanoparticles for therapeutic purpose [165]. Sodium alginate/PVA-titanium dioxide (TiO_2) based wound healing patches were synthesised. TiO_2 provides characters such as biocompatibility, no toxicity, antibacterial and antifungal etc. [165]. TiO_2 NPs and curcumin were incorporated into the polymeric patches and tested for antibacterial activity against *Bacillus subtilis*, *Klebsiella pneumoniae*, *Staphylococcus aureus*, and *Escherichia coli*, as well as antifungal activity against *Candida albicans* and *Aspergillus niger* [92]. This provides evidence that both TiO_2 nanoparticles and curcumin incorporated SA/PVA patches can be better used in wound healing [165].

Non-biodegradable, opaque and occlusive and low swelling capacity are the vital obstacles of few commercially available wound dressings materials used for clinical performance. Thus to overcome and improve such drawbacks, a novel biodegradable wound dressing material was prepared by using alginate membrane and polycaprolactone (PCL) nanoparticles loaded with curcumin [164]. Curcumin was employed as a model drug due to its important properties in wound healing, including antimicrobial, antifungal, and anti-inflammatory effects. Both, *in-vitro* and *in-vivo* trials were conducted to assess the possible use of this wound dressing material. The novel membrane had a wide range of functional properties that made it suitable for use as a synthetic skin substitute, including a high capacity for swelling and adherence to the skin, evidence of pores to control the loss of trans-epidermal water, clarity for wound monitoring, a high capacity for swelling, controlled drug release and effective adherence to the skin. The use of nanocarriers aids the drug's permeation through various skin layers, resolving curcumin's solubility issues. It also claimed that the clinical implementation of this method would cover large areas of mixed first- and second-degree wounds without the need for removal, reducing patient pain and the risk of changing the formation of the new epithelium tissues [164].

Recently biocompatible polymers are widely used in wound care systems especially for burn wounds as well as skin damages. Polymers such as polyvinyl alcohol and chitosan have been shown to be biocompatible with low toxicity have make them ideal for treating injuries with limited side effects [108]. Curcumin, a key component of turmeric, has anti-inflammatory and antimicrobial effects, but its bioavailability is extremely poor. Curcumin's bioavailability was increased exponentially after it was converted to its nano form, and allowing it to play an important role in wound healing. Considering the biopolymer and dynamic nature of curcumin, polymeric patches were prepared from Polyvinyl alcohol and chitosan with nano-curcumin for wound healing purpose. Slow vaporisation was used to successfully synthesise nano-curcumin, which was then integrated into polyvinyl alcohol and chitosan and obtained as a PVA/Chi/Cur patch using the gel casting process. Different characterisation methods were used to assess the patch: wwelling rate, evaporation rate, blood effect, cell biocompatibility, and antibacterial activity were all investigated. The patch was screened for antibacterial activity against common bacteria present on the wound site (*Escherichia coli*, *Pseudomonas aeruginosa*, *Staphylococcus aureus*, and *Bacillus subtilis*). The study was completed by putting the patch wound healing efficiency to the test on Albino Westar rats for 16 days. The tests were successful and results of cell line experiments and the MTT-assay showed that the PVA/Chi/Cur patch improved cell proliferation. Its antibacterial activity against four major bacterial strains present in wound sites, as well as its water retention, makes it an ideal material for wound healing. Its superiority over commercial ointment was demonstrated in *in-vivo* trials. This procedure for epidermal wounds decreases the number of times the patch has to be replaced and speeds up wound healing [108].

The β-cyclodextrin (CD) is generally used as a good stabilising and solubilising agent for preparation of different pharmaceutical products. It also enhances the water solubility of curcumin by forming β-cyclodextrin (CD)–Cur complex [163]. To facilitate cutaneous wound healing, a composite prepared from curcumin (Cur), chitosan–alginate (CA) and β-cyclodextrin (CD) [163] by adding curcumin to the ring-shaped β-cyclodextrin (CD) to form a β-CD–Cur inclusion complex (CD-Cur) and tested in a skin model to shows higher skin permeability than free Cur [174–176]. These findings indicate that Cur-CD can be an useful component for cutaneous wound healing materials. Animal studies with cutaneous wounds in rats using CA-CD-Cur showed accelerated and better wound closure rates, good histopathological results, and lower SOD, lipid peroxidation, pI3K, and pAkt levels in comparison to other material. Thus, CA-CD-Cur can facilitate cutaneous wound dressing that facilitate faster wound healing process [163].

Another alginate based hydrogel system was developed using curcumin-β-cyclodextrin based inclusion complex for wound healing purpose [162]. The fabricated curcumin-β-cyclodextrin inclusion complex loaded sodium alginate/chitosan (CMx-loaded SA/CS) bilayer hydrogels was tested as a better wound healing materials. The improve materials with high $CaCl_2$ content has high tensile strength, percent elongation at split and Young's modulus. The release characteristics of CMx from all hydrogels exhibited a similar pattern of release. Furthermore, the CMx-loaded SA/CS bilayer hydrogels inhibited Gram-negative (*Escherichia coli*) as well as Gram-positive bacteria (*Staphylococcus aureus*). Finally, NCTC clone 929 and NHDF (normal human dermal fibroblast) cells were found to be non-toxic to all bilayer hydrogels. Therefore, these CMx loaded SA/CS bilayer hydrogels had the potency for wound healing and can be used as wound dressing materials [162].

As newer technologies are approaching for development of proper wound care management systems, elastin based biomimetics are one of the trending topic for current material development. This concept contributes for the preparation and characterisation of series of cross-linked films based on the combination of an elastin-derived biomimetic polypeptide [Human Elastin-Like Polypeptide (HELP)] with alginate (ALG) to obtain a composite with enhanced properties [161]. There are a few examples of such elastin-like composites produced to date by combining alginate and HELP to tune the final properties of the resulting material, modulating the delivery of curcumin as a natural molecule used as an antioxidant compound. The existence of HELP in the composite was shown to be useful in controlling the release of the model compound curcumin, resulting in a high antioxidant activity of the material as well as maintaining and improving the final material's cytocompatibility. More research is required to assess the *in-vivo* behaviour of this composite material. However, the current findings showed that combining alginate with HELP to create customizable platforms for drug delivery, wound healing, and tissue regeneration is effective. Finally, HELP-based proteins can be easily customised by molecular fusion of exogenous domains to prepare dynamic biopolymers. These HELP fusion proteins may be used in the future to provide additional flexibility to final composite materials for therapeutic use [161].

Graphene oxide is one of the most preferable material which have characteristics like biocompatibility, greater surface area, high mechanical strength and also antimicrobial activity. *Gymnema sylvestre* is commonly known as "Gumar" plant product also having anti-inflammatory, antimicrobial properties [160]. Fabrication of a novel scaffold was done consisting of curcumin and *Gymnema sylvestre* incorporated graphene oxide polyhydroxy butyrate-sodium alginate (GO-PHB-SA-CUR and GS) composite as an extracellular matrix platform to improve wound healing in normal and diabetic wounds [160]. Curcumin and *Gymnema sylvestre* were integrated into the novel graphene oxide-polyhydroxybutyrate sodium alginate composite scaffold prepared by solution casting to enhance wound healing in regular and diabetic wounds for better tissue regeneration use. The particle size, surface charge (zeta potential), crystalline nature (XRD), and morphology (FE-SEM) content of the GO-PHB-SA-CUR and GS composite were all assessed. The presence of GO-PHB-SA improves wound closure and increases cell viability in both wounded and diabetic wound cells without causing cytotoxicity. The manufacturing of mesoporous composites improves fibroblast cell viability. The composite made from GO-PHB-SA-CUR and GS has been tested as a wound dressing material by clinical trials to validate its future use. *In-vitro* tests with normal and diabetic fibroblast cells indicated that the GO-PHB-SA-CUR and GS composite had strong biocompatibility in terms of increased wound cell migration. As a result the GO-PHB-SA-CUR and GS composite could help regular and diabetic wounds heal faster. According to the findings the GO-PHB-SA-CUR and GS composite scaffold appears to be a promising candidate for a new wound dressing material that is both reliable and affordable. More successful clinical trials of this material will help millions of diabetic patients with improve wound healing capacity and the quality of life they are going through [160].

Multifunctional biopolymer composites comprising mechanically disintegrated bacterial cellulose, alginate, gelatine and curcumin plasticized with glycerol (BCAGG-C) were also successfully made through a simple, naive, cost-supportive mechanical blending and casting method [159]. The composites had a well distributed structure, according to FE-SEM pictures. The water touch angles ranges from 50 to 70 degrees and its permeability value was 300–800 $g/m^2/24$ hrs, which was equivalent to commercially available dressing products. When the fil was immersion in phosphate buffer solution (PBS) and artificial saliva, no curcumin was released from the films and the fluid uptakes were in the range of 100–700%. Mechanical properties revealed that BCAGG-C films had sufficient strength and versatility to be used as wound dressings. The stretchable film provides adequate stiffness and long-term deformation. The skin was tightly adhered by hydrated films. Under artificial saliva medium, the *in-vitro* muco-adhesion time was found to be in the range of 0.5–6 hrs using porcine mucosa as a model membrane. Despite being trapped within the biopolymer matrix composite, curcumin could possess useful biological activities. The curcumin-loaded films had successful antibacterial activity against *E. coli* and *S. aureus*. Human keratinocytes and gingival fibroblasts were not harmed by the films, but oral cancer cells displayed potent anticancer activity. As a result, these curcumin-loaded films can be used as leave-on skin applications. These inventive films can be further adapted to meet the needs of local topical patches for wound care, periodontitis, and also for mouth cancer [159].

Taking almost the similar components such as biocompatible biopolymers, sodium alginate and gelatine microfibers are prepared via extrusion-gelation into a physical crosslinking solution [158]. Curcumin, which is an ancient natural bioactive wound healing agent, was loaded into the fibbers. Biopolymers including sodium alginate and gelatine were used to make curcumin-loaded composite microfibers and blank microfibers. The different concentrations of sodium alginate and gelatine in the formulation batches were coded as A1G9-A10G0. The ATR-FTIR was used to describe the molecular transitions inside the composite microfibers, which was then confirmed using molecular dynamic analysis. The mechanical properties such as tensile strength and elongation-at-break (extensibility) were varying between 1.08 ± 0.01 to 3.53 ± 0.41 N/mm^2 and 3.89 ± 0.18% to 0.61 ± 0.03% respectively. The microfibers' formation and fabrication were verified by morphological examination. Physical evaluations, such as matrix degradation and entrapment performance, were also carried out to provide a comparative account of the various formulations. The water uptake ability of the blank and curcumin-loaded composite fibres is found to be 30.77 ± 2.17 to 100.00 ± 5.99 and 22.34 ± 1.11 to 56.34 ± 4.68, respectively. The cumulative release of composite microfibers was 85% in 72 hours, which demonstrating the composite fibres' long-term release capacity. The drug was released in an unusual (non-Fickian) pattern, implying the importance of degradation and diffusion. In an *in-vivo* full-thickness cutaneous wound model, the composite microfibers had higher degree of contraction (96.89 ± 3.76%) than the commercially available lead products such as Vicco turmeric cream. It can be concluded that natural, alginate–gelatine–curcumin composite has the potency to be explored as a cost-effective wound healing product [158].

13. Conclusions

Despite of rapid and extensive researches on curcumin as a natural ingredient with alginate for formulation of wound dressing materials for proper wound healing, till today, no such curcumin-alginate based commercial products are available in the market for pharmaceutical use. Recently, many patents have been filled and approved for preparing curcumin based biopolymers useful for wound healing materials, but in practical, the use of the same is in clinical stage. As stated in the text, curcumin has broad range of pharmacological activities, its antioxidant and anti-inflammatory properties, in particular, suggest that it may be very effective and useful in wound dressing [154]. As a best Indian herbal medicine over the years, curcumin topical formulations, including nano-architectures, have been developed and tested for enhancing curcumin's wound-healing activity. The key reason for choosing the topical nano-formulation of curcumin is that it provides better solubility, bioavailability and sustained release of curcumin in an active form, all of which are essential for delivering a consistent dose of the drug for a long period which further help in wound healing. Curcumin's perfect dose is critical for

DOI: http://dx.doi.org/10.5772/intechopen.98830

a variety of reasons, but most importantly, its complex function in the inflammatory response in wound healing must be discussed before further clinical progress. As summarised in this chapter various topical formulations of curcumin are being developed with the aim of delivering curcumin to the wounded site in a sustained manner to enhance therapeutic outcomes. Nonetheless, the molecular mechanisms underlying its ability to control chronic inflammation and influence the genetic, cellular wound environment are expected to be studied in depth. While current research on various topical formulations of curcumin appears promising, the majority of published evidence is based on in vitro and *in-vivo* studies, and clinical trials are still required. Thus, in the near future, experiments on human clinical trials should provide answers to concerns about the safety of different topical formulations based on curcumin-alginate in biological systems, as well as their therapeutic wound-healing efficacy.

Author details

Sulata Sahu[1] and Bairagi C. Mallick[2*]

1 Department of Zoology, Ravenshaw University, Cuttack, Odisha, India

2 Department of Chemistry, Ravenshaw University, Cuttack, Odisha, India

*Address all correspondence to: bcmallick@ravenshawuniversity.ac.in

References

[1] Aderibigbe, B.A. and Buyana, B., Alginate in Wound Dressings. *Pharmaceutics.* 2018, 10(42), 2-19. doi:10.3390/pharmaceutics10020042

[2] Cullum, N., Buckley, H., Dumville, J., Hall, J., Lamb, K., Madden, M., Morley, R., O'Meara, S., Goncalves, P.R., Soares, M., Wounds research for patient benefit: a 5-year programme of research. *NIHR Journals Library.* 2016. doi:10.3310/pgfar04130

[3] Frykberg, R.G., and Banks, J., Challenges in the Treatment of Chronic Wounds. *Adv Wound Care (New Rochelle)*2015, 4, 560-582.doi: 10.1089/wound.2015.0635

[4] Schmidtchen, A., Pang, C. Ni, G.,Sonnergren, H., Car, J. Järbrink, K., Bajpai, R., The humanistic and economic burden of chronic wounds: A protocol for a systematic review. *Syst. Rev*. 2017,6(15),1-7. doi:10.1186/s13643-016-0400-8

[5] Flanagan, M. Barriers to the Implementation of Best Practice in Wound Care. *Wounds UK.* Available online: http://www.wounds-uk.com/pdf/content_9030.pdf

[6] Ulery, B. D., Nair, L. S., Laurencin, C.T., Biomedical applications of biodegradable polymers. *J. Polym. Sci. Pol. Phys*. 2011, 49(12), 832-864. doi:10.1002%2Fpolb.22259

[7] Yadav, P., Yadav, H., Shah, V.G., Shah, G., Dhaka, G., Biomedical biopolymers, their origin and evolution in biomedical sciences: A systematic review. *J. Clin. Diagn. Res*. 2015, 9(9), ZE21–ZE25 doi: 10.7860/JCDR/2015/13907.6565

[8] Sudarsan, S., Franklin, D.S., Guhanathan, S., Imbibed salts and pH-responsive behaviours of sodium-alginate based eco-friendly biopolymeric hydrogels:A solventless approach. *MMAIJ.* 2015, 11(1), 24-29.

[9] Varaprasad, K., Jayaramudu, T., Kanikireddy, V., Toro, C., Sadiku, ER., Alginate-based composite materials for wound dressing application: a mini review. *CarbohydrPolym.* 2020, 236:116025. doi: 10.1016/j.carbpol.2020.116025

[10] Sweeney, IR., Miraftab, M., Collyer, G., Absorbent alginate fibres modified with hydrolysed chitosan for wound care dressings—II. Pilot scale development. *CarbohydrPolym.* 2014,102, 920-927. doi: 10.1016/j.carbpol.2013.10.053

[11] Gupta, A., Kowalczuk, M., Heaselgrave, W., Britland, ST., Martin, C., Radecka, I., The production and application of hydrogels for wound management: a review. *Eur Polym J.* 2019,111,134-151.

[12] Kurakula, M., Rao, G. S. N. K., Kiran, V., Hasnain, M. S., and Nayak, A.K., Alginate-based hydrogel systems for drug releasing in wound healing. Alginates in Drug Delivery. *Elsevier Inc.* 2020, Book Chapter 13, 323-358

[13] Sood, A., Granick, MS., Tomaselli, NL., Wound dressings and comparative effectiveness data. *Adv Wound Care.* 2014,3(8),511-529. doi: 10.1089/wound.2012.0401

[14] Malafaya, PB., Silva, GA., Reis, RL. Natural-origin polymers as carriers and scaffolds for biomolecules and cell delivery in tissue engineering applications. *Adv Drug Deliv Rev.* 2007,59(4-5),207-233. doi: 10.1016/j.addr.2007.03.012.

[15] Dowling, MB., Chaturvedi, A., MacIntire, IC., Javvaji, V., Gustin, J., Raghvan, SR., Scalea, TM., Narayan, M.,Determination of efficacy of a novel

alginate dressing in a lethal arterial injury model in swine. *Injury.* 2016, 47(10), 2105-2109. doi: 10.1016/j.injury.2016.05.003.

[16] Matricardi, P., Meo, C. D., Coviello, T., Alhaique, F., Recent advances and perspectives on coated alginate microspheres for modified drug delivery. *Expert Opin Drug Deliv.* 2008,5(4),417-425. doi:10.1517/17425247.5.4.417

[17] Gombotz, WR., Wee, SF., Protein release from alginate matrices. *Adv Drug Delivery Rev.* 1998;31:267-85. doi: 10.1016/s0169-409x(97)00124-5.

[18] Lee, K. Y., Mooneya, D.J. Alginate: Properties and biomedical applications Progress in *Polymer Science.* 2012, 37(1), 106– 126. doi: 10.1016/j.progpolymsci.2011.06.003.

[19] Dhargalkar, V. K., Kavlekar, D., Seaweeds: A field manual. National Institute of Oceanography, Dona Paula, Goa. 2004, 1-42.

[20] Milani, J., Maleki, G., Hydrocolloids in food industry. Food industrial processes – Methods and equipment. *In Tech, Rijeka*, 2012, Book Chapter 2 17-21. (2012).

[21] Pereira, L. A review of the nutrient composition of selected edible seaweeds. Seaweed: Ecology, nutrient composition and medicinal uses. *Nova Science Publishers, New York*, 2011, Book Chapter 1, 15-47.

[22] Remminghorst, U., Rehm, B. H. A., Bacterial alginates: From biosynthesis to applications. *Biotechnology Letters*, 2006, 28(21), 1701-1712. doi: 10.1007/s10529-006-9156-x

[23] Draget, K. I. Alginates. Hand book of hydrocolloids,Woodhead Publishing Series in *Food Science, Technology and Nutrition*. 2009, Book Chapter 29, 807-82

[24] McHugh, D. J.: A guide to seaweed industry. FAO Fisheries and Aquaculture Department, Rome, 2003

[25] Pawar, S.N., Edgar, K. J.Alginate derivatization: A review of chemistry, properties and applications. *Biomaterials.* 2012, 33, 3279-3305 doi:10.1016/j.biomaterials.2012.01.00

[26] Fischer, FG, Dorfel, H., Die polyuronsauren der braunalgen-(kohlenhydrate der algen-I). *Z. Physiol. Chem.* 1955,302(4-6),186-203

[27] Haug, A., Fractionation of alginic acid. *Acta Chem Scand.* 1959;13,601-3

[28] Tonnesen, H. H., Karlsen, J., Alginate in drug delivery systems. *Drug Dev Ind Pharm.* 2002,28(60),621-30. doi: 10.1081/ddc-120003853

[29] Qin, Y. Alginate fibres: an overview of the production processes and applications in wound management. *Polym Int.* 2008,57(2),171-80. doi:10.1002/pi.2296

[30] George, M., Abraham, TE., Polyionic hydrocolloids for the intestinal delivery of protein drugs. *J Control Release.* 2006,114(1),1-14. doi: 10.1016/j.jconrel.2006.04.017.

[31] Hay, ID., Rehman, ZU., Ghafoor, A., Rehm, BHA., Bacterial biosynthesis of alginates. *J Chem TechnolBiotechnol.* 2010,85(6),752-759. doi:10.1002/jctb.2372

[32] Rinaudo, M. On the abnormal exponents a_η and a_D in Mark–Houwink type equations for wormlike chain polysaccharides. *Polym Bull.* 1992,27, 585-589. doi:10.1007/BF00300608

[33] LeRoux, MA., Guilak, F., Setton, LA., Compressive and shear properties of alginate gel: effects of sodium ions and alginate concentration. *J Biomed Mater Res.* 1999;47(1),46-53.

doi: 10.1002/(SICI)1097-4636 (199910)47:13.0.CO;2-N ·

[34] Kong, HJ., Lee, KY., Mooney, DJ., Decoupling the dependence of rheological/mechanical properties of hydrogels from solids concentration. *Polymer.* 2002,43(23),6239-6246. doi:10.1016/S0032-3861(02)00559-1

[35] Franklin, MJ., Douthit, SA., McClure, MA., Evidence that the algI/ algJ gene cassette, required for O acetylation of *Pseudomonas aeruginosa* alginate, evolved by lateral gene transfer. *J Bacteriol.* 2004,186(14),4759-4773. doi: 10.1128/JB.186.14.4759-4773. 2004.

[36] Franklin, MJ., Ohman, DE., Mutant analysis and cellular localization of the AlgI, AlgJ, and AlgF proteins required for O acetylation of alginate in *Pseudomonas aeruginosa*. *J Bacteriol.* 2002,184(11),3000-3007. doi: 10.1128/ JB.184.11.3000-3007.2002

[37] Chitnis, CE., Ohman, DE., Cloning of *Pseudomonas aeruginosa* algG, which controls alginate structure. *J Bacteriol.* 1990,172(6),2894-2900. doi: 10.1128/ jb.172.6.2894-2900.1990.

[38] Franklin, MJ., Chitnis, CE., Gacesa, P., Sonesson, A., White, DC., Ohman, DE., Pseudomonas aeruginosa AlgG is a polymer level alginate C5-mannuronan epimerase. *J Bacterio.l* 1994,176(7),1821-1830. doi: 10.1128/jb.176.7.1821-1830. 1994.

[39] Franklin, MJ., Ohman, DE., Identification of algF in the alginate biosynthetic gene cluster of *Pseudomonas aeruginosa* which is required for alginate acetylation. J Bacteriol 1993;175(16):5057e65. doi: 10.1128/jb.175.16.5057-5065.1993

[40] Gimmestad, M., Sletta, H., Ertesvag, H., Bakkevig, K., Jain, S., Suh, SJ., The *Pseudomonas fluorescens* AlgG protein, but not its mannuronan C-5- epimerase activity, is needed for alginate polymer formation. *J Bacteriol* 2003,185(12),3515-3523. doi: 10.1128/ jb.185.12.3515-3523.2003.

[41] Goh, CH., Heng, PWS., Chan, LW., Alginates as a useful natural polymer for microencapsulation and therapeutic applications. *CarbohydrPolym.* 2012;88(1),1-12. doi:10.1016/j. carbpol.2011.11.012.

[42] Hasnain, MS., Ahmad, SA., Chaudhary, N., Hoda, MN., Nayak, AK., Biodegradable polymer matrix nanocomposites for bone tissue engineering. *Elsevier Inc*; 2019, Book Chapter 1, 1-37.

[43] Smidsrod, O., Skjaok-Braek, G., Alginate as immobilization matrix for cells. *Biotechnol.*1990,8,71-8. doi:10. 1016/0167-7799(90)90139-

[44] Nayak, AK., Ahmed, SA., Tabish, M., Hasnain, MS., Natural polysaccharides in tissue engineering application. Natural polysaccharides in drug delivery and biomedical applications. Academic Press, *Elsevier Inc*. 2019, Book Chapter 23, 531-548. doi:10.1016/B978-0-12-817055-7. 00023-6

[45] Pal, D., Nayak, AK., Development, optimization and anti-diabetic activity of gliclazide loaded alginate-methyl cellulose mucoadhesive microcapsules. *AAPS Pharm Sci Tech.* 2011,12(4),1431-1441. doi: 10.1208/s12249-011-9709-8.

[46] Bera, H., Kandukuri, SG., Nayak, AK., Boddupalli, S., Alginate-sterculia gum gel-coated oil-entrapped alginate beads for gastroretentive risperidone delivery. *Carbohydr Polym.* 2015,120,74-84. doi: 10.1016/j.carbpol.2014.12.009.

[47] Hasnain, MS., Rishishwar, P., Rishishwar, S., Ali, S., Nayak, AK., Isolation and characterization of Linumusitatisimum polysaccharide to prepare mucoadhesive beads of

diclofenac sodium. *Int. J. Biol. Macromol.* 2018,116,162-72. doi: 10.1016/j.ijbiomac.2018.04.151.

[48] Pal, D., Saha, S., Nayak, AK., Hasnain, MS., Marine-derived polysaccharides: pharmaceutical applications. Natural polymers for pharmaceutical applications, vol. II: Marine and microbiologically derived polymers. Palm Bay, FL: *Apple. Academic Press.* 2019, 1-36. https://doi.org/10.1201/9780429328299

[49] Rath Adhikari, SN., Nayak, BS., Nayak, AK., Mohanty, B., Formulation and evaluation of buccal patches for delivery of atenolol. *AAPS Pharm. Sci. Tech.* 2010,11 (3),1034-1044. doi: 10.1208/s12249-010-9459-z

[50] Yadav, M., Ahmadi, Y., Alginates: source, chemistry, and properties.: Alginate: versatile polymer in biomedical applications and therapeutics. Palm Bay, FL: *Apple Academic Press*. 2019, 1-24.

[51] Brigger, I., Dubernet, C., Couvreur, P., Nanoparticles in cancer therapy and diagnosis. *Adv. Drug Deliv. Rev*. 2012, 64, 24-36. doi:10.1016/j.addr.2012.09.006

[52] Hamidi, M., Azadi, A., Rafiei, P., Hydrogel nanoparticles in drug delivery. *Adv Drug Deliv Rev*. 2008, 60(15), 1638-1649. doi:10.1016/j.addr.2008.08.00

[53] Lambert, G., Fattal, E., Couvreur, P., Nanoparticulate systems for the delivery of antisense oligonucleotides. *Adv. Drug Deliv. Rev*. 2001, 47(1), 99-112. doi: 10.1016/s0169-409x(00)00116-2.

[54] Lertsutthiwong, P., Rojsitthisak, P., Chitosan-alginate nanocapsules for encapsulation of turmeric oil. *Pharmazie.* 2011, 66(12), 911-915.

[55] Soppimath, KS., Aminabhavi, TM., Kulkarni, AR., Rudzinski, WE., Biodegradable polymeric nanoparticles as drug delivery devices. *J. Control Release.* 2001, 70(1-2), 1-20. doi: 10.1016/s0168-3659(00)00339-4.

[56] Severino, P., Silva, C.F., Andrade, L.N., Oliveira, D.L., Campos, J. and Souto, E.B., Alginate Nanoparticles for Drug Delivery and Targeting. *Current Pharmaceutical Design.* 2019, 25(11), 1312-1334. doi: 10.2174/1381612825666190425163424.

[57] Fernando, I.P.S., Lee, W.W., Han, E.J., Ahn, G., Alginate-based nanomaterials: fabrication techniques, properties, and applications. *Chemical Engineering Journal.* 2019,123823. doi:10.1016/j.cej.2019.123823.

[58] Hoque, J., and Haldar, J., Direct synthesis of dextran-based antibacterial hydrogels for extended release of biocides and eradication of topical biofilms. *ACS Applied Materials and Interfaces.* 2017, 9(19), 15975-15985. doi: 10.1021/acsami.7b03208.

[59] Lazarus, GS., Cooper, DM., Knighton, DR., Margolis, DJ., Percoraro, ER., Rodeheaver, G., Robson, MC., Definitions and guidelines for assessment of wounds and evaluation of healing. *Arch. Dermatol.* 1994,130(4), 489-93.

[60] Alam, P. Design considerations in the development of wound healing bionanomaterials. Nanostructures for novel therapy. *Elsevier Inc*. 2017, Book Chapter 13 343-73. doi:10.1016/B978-0-323-46142-9.00013-X

[61] Baranoski, S., Ayello, EA., Langemo, DK., Wound assessment. Wound care essentials: practice principles. Philadelphia, PA: *Lippincott Williams &Wilkins*. 2008, Book Chapter 6,79-90.

[62] Chaudhari, AA., Vig, K., Baganizi, DR., Sahu, R., Dixit, S., Dennis, V.,. Future prospects for scaffolding methods and biomaterials in skin tissue engineering: a review. *Int. J. Mol. Sci.*

2016, 17(12),1974. doi: 10.3390/ijms17 121974.

[63] Dhivya, S., Padma, VV., Santhini, E., Wound dressingsa review. *Biomed.* 2015, 5(4), 22. doi: 10.7603/ s40681-015-0022-9

[64] Parani, M., Lokhande, G., Singh, A., Gaharwar, A. K., Engineered nanomaterials for infection control and healing acute and chronic wounds. *ACS Applied Materials and Interfaces.* 2016, 8(16), 10049-10069. doi:10.1021/ acsami.6b0029

[65] Stefanov, I., Pérez-Rafael, S., Hoyo, J., Cailloux, J., Santana Pérez, O. O., Hinojosa- Caballero, D., Tzanov, T., Multifunctional enzymatically generated hydrogels for chronic wound application. *Biomacromolecules.* 2017, 18(5), 1544-1555. doi:10.1021/acs. biomac.7b00111

[66] Eming, SA., Martin, P., Tomic-Canic, M., Wound repair and regeneration: mechanisms, signaling and translation. *Sci. Transl. Med.* 2014, 6(265). doi: 10.1126/scitranslmed. 3009337

[67] Bolton, L., van Rijswijk, L., Wound dressings: meeting clinical and biological needs. Dermatol Nurs. 1991,3(3),146-161.

[68] Krasner, D., Sennedy, KL., Rolstad, BS., Roma, AW., The ABCs of wound care dressings. *Ostomy Wound Manage* 1993, 39(8), 68-72.

[69] Ferreira, MC., Tuma, Jr P., Carvalho, VF., Kamamoto, F., Complex wounds. *Clinics.* 2006, 61(6), 571-578. doi:10.1590/S1807-59322006000 600014

[70] De Roche, R., Luscher, N. J., Debrunner, H. U., Fischer, R., Epidemiological data and costs of burn injuries in workers in Switzerland: an argument for immediate treatment in burn centres. *Burns.* 1994,20(1),58-60. doi: 10.1016/0305-4179(94)90108-2.

[71] Thomas, S., Assessment and management of wound exudates. *J Wound Care.* 1997,6 (7), 327-30.

[72] Sezer, A.D., Cevher, E., Biopolymers as wound healing materials: Challenges and new strategies. In Biomaterials Applications for Nanomedicine; Pignatello, R., Ed.; InTech: Rijeka, Croatia, 2011, Book Chapter 19,383-414.

[73] Zahedi, P., Rezaeian, I., Ranaei-Siadat, S.O., Jafari, S.H., Supaphol, P., A review on wound dressings with an emphasis on electrospun nanofibrous polymeric bandages. *Polym. Adv. Technol*. 2010, 21, 77-95. https://doi. org/10.1002/pat.1625

[74] Oro, F.B., Sikka, R.S., Wolters, B., Graver, R., Boyd, J.L., Nelson, B.,Swiontkowski, M.F., Autograft versus allograft: An economic cost comparison of anterior cruciate ligament reconstruction. *Arthroscopy*. 2011, 27(9), 1219-1225. doi: 10.1016/j.arthro. 2011.04.008.

[75] Sheikh, Z., Hamdan, N., Ikeda, Y., Grynpas, M., Ganss, B., Glogauer, M., Natural graft tissues and synthetic biomaterials for periodontal and alveolar bone reconstructive applications: A review. *Biomater. Res.* 2017, 21, 9. 24. doi: 10.1186/ s40824-017-0095-5.

[76] Guo, S.A., DiPietro, L.A., Factors affecting wound healing. *J. Dental Res.* 2010, 89(3), 219-229. doi: 10.1177/0022034509359125

[77] Sanchez, J. L., Pereperez, S. B., Bastida, J. L., Martinez, M. M., Cost-utility analysis applied to the treatment of burn patients in a specialized canter. *Arch. Surg.* 2007,142 (1), 50-7. doi: 10.1001/archsurg.142.1.50.

[78] Zuliani-Alvarez, L., Midwood, K.S., Fibrinogen-related proteins in tissue

repair: How a unique domain with a common structure controls diverse aspects of wound healing. Adv. *Wound Care*. 2015, 4(5), 273-285. doi: 10.1089/wound.2014.0599.

[79] Sun, B.K., Siprashvili, Z., Khavari, P.A., Advances in skin grafting and treatment of cutaneous wounds. *Science.* 2014, 346(6212), 941-945. doi: 10.1126/science.1253836.

[80] Aller, M.A., Arias, J.I., Arraez-Aybar, L.A., Gilsanz, C., Arias, J., Wound healing reaction: A switch from gestation to senescence. *World J. Exp. Med.* 2014, 4(2), 16-26. doi: 10.5493/wjem.v4.i2.16.

[81] Khor, E., Lim, LY., Implantable applications of chitin and chitosan. *Biomaterials*. 2003, 24(13), 2339-2349. https://doi.org/10.1016/S0142-9612(03)00026-7

[82] Loke, W. K., Lau, S. K., Yong, L. L., Khor, E., Sum, C. K., Wound dressing with sustained anti-microbial capability. *J Biomed Mater Res*. 2000, 53(1):8-17.doi: 10.1002/(sici)1097-4636(2000)53:1<8::aid-jbm2>3.0.co;2-3

[83] Ma, Y., Xin, L., Tan, H., Chitosan membrane dressings toughened by glycerol to load antibacterial drugs for wound healing. *Mat Sci Eng C Mater*. 2017 81:522-531. doi: 10.1016/j.msec.2017.08.052.

[84] Jayakumar, R., Prabaharan, M., Kumar, PTS., Nair, SV., Tamura, H. Biomaterials based on chitin and chitosan in wound dressing applications. *Biotechnol Adv*. 2011, 29(3),322-337. doi: 10.1016/j.biotechadv.2011.01.005.

[85] Xiea, H., Chena, X., Shena, X., Hea, Y., Chena, W., Luoa, Q., Geb, W., Yuana, W., Tanga, X., Houa, D., Jianga, D., Wanga, Q., Liua, Y., Liua, Q., Lia, K., The Preparation of chitosan-collagen-alginate composite dressing and itspromoting effects on wound healing. *Int J. Biol. Macromol*. 2018 107(Pta A):93-104 doi: 10.1016/j.ijbiomac.2017.08.142.

[86] Michalska-Sionkowska, M., Walczak, M., Sionkowska, A. Antimicrobial activity of collagen material with thymol addition for potential application as wound dressing. *Polym Test. 2017* 63:360-366. https://doi.org/10.1016/j.polymertesting.2017.08.03

[87] Pal, S., Nisi, R., Stoppa, M., Licciulli, A., Silver-functionalized bacterial cellulose as antibacterial membrane for wound-healing applications. *ACS Omega*. 2017, 2(7),3632-3639 https://doi.org/10.1021/acsomega.7b00442

[88] You, C., Li, Q., Wang, X. Wu, P., Ho AK.,Silver nanoparticle loaded collagen/chitosan scaffolds promote wound healing via regulating fibroblast migration and macrophage activation. *Sci Rep*. 2017, 7(1):10489 doi: 10.1038/s41598-017-10481-0.

[89] Dowling, MB., Chaturvedi, A., MacIntire, IC., Determination of efficacy of a novel alginate dressing in a lethal arterial injury model in swine. *Injury*. 201647(10),2105-2109. doi: 10.1016/j.injury.2016.05.003.

[90] Ye, J., Shi, X., Chen, X. Chitosan-modified, collagen- based biomimetic nanofibrous membranes as selective cell adhering wound dressings in the treatment of chemically burned corneas. *J Mater Chem B*, 2014 27:4226-4236 https://doi.org/10.1039/C3TB21845G

[91] Baniasadi, M., Minary-Jolandan, M., Alginate–collagen fibril composite hydrogel. *Materials (Basel).* 2015, 8(2),799-814 doi: 10.3390/ma8020799

[92] Lee, BR., Hwang, JW., Choi, YY., Wong, SF., Hwang, YH., Lee, DY., Lee, SH., In situ formation and

collagen–alginate composite encapsulation of pancreatic islet spheroids. Biomaterials 33(3):837-845. doi: 10.1016/j.biomaterials.2011.10.014.

[93] Lee, H-j., Ahn S-H., Kim GH., Three-dimensional collagen/ alginate hybrid scaffolds functionalized with a drug delivery system (DDS) for bone tissue regeneration. *Chem Mater*. 2011 24(5),881-891 https://doi.org/10.1021/cm200733s

[94] Zhang, H., Peng, M. , Cheng, T. , Zhao, P., Qiu, L. , Zhou, J. , Lu, G.,& Chen, J., Silver nanoparticles-doped collagen–alginate antimicrobial biocomposite as potential wound dressing. *J Mater Sci*. 2018 ,53,14944-14952 doi:10.1007/s10853-018-2710-9

[95] Azarniya, A., Tamjid, E., Eslahi, N., &Simchi, A., Modification of bacterial cellulose/keratin nanofibrous mats by a tragacanth gum-conjugated hydrogel for wound healing. *International Journal of Biological Macromolecules*. 2019, 134, 280-289. doi:10.1016/j.ijbiomac.2019.05.023

[96] Archana, D., Singh, B.K., Dutta, J., Dutta, P.K., Chitosan-PVP-nano silver oxidewound dressing: in vitro and in vivo evaluation. *Int. J. Biol. Macromol*. 2015, 73,49-57. doi: 10.1016/j.ijbiomac.2014.10.055

[97] Kumar, S., Dutta, ., Dutta, PK., Preparation and characterization of N-heterocyclic chitosan derivative based gels for biomedical applications, *Int.J. Biol. Macromol*. 2009, 45(4), 330-337. doi:10.1016/j.ijbiomac.2009.08.002

[98] Behera, SS., Das, U., Kumar, A., Bissoyi, A., Singh, A.K., Chitosan/ TiO2compositemembrane improves proliferation and survival of L929 fibroblast cells:application in wound dressing and skin regeneration, *Int. J. Biol. Macromol.* 2017, 98329-340. doi: 10.1016/j.ijbiomac.2017.02.017.

[99] Akbik, D., Ghadiri, M., Chrzanowski, W., Rohanizadeh, R., Curcumin as a wound healing agent. *Life Sci.* 2014, 116(1), 1–7. doi: 10.1016/j.lfs.2014.08.016.

[100] Gupta, A., Briffa, S.M., Swingler, S., Gibson, H., Kannappan, V., Adamus, G., Kowalczuk, M., Martin, C., and Radecka, I., Synthesis of Silver Nanoparticles Using Curcumin-Cyclodextrins Loaded into Bacterial Cellulose-Based Hydrogels for Wound Dressing Applications. *Biomacromolecules.* 2020, 21(5), 1802–1811. doi: 10.1021/acs.biomac.9b01724.

[101] Gupta, A. Keddie, D. J. Kannappan, V. Gibson, H. Khalil, I.R. Kowalczuk, M. Martin, C. Shuai, X. Radecka, I., Production and characterisation of bacterial cellulose hydrogels loaded with curcumin encapsulated in cyclodextrins as wound dressings. *Eur. Polym. J.* 2019,118, 437–450. doi:10.1016/j.eurpolymj.2019.06.018

[102] Gong, CY., Wu, Q., Wang, Y., Zhang, DD., Luo, F, Zhao, X., Wei, Y., Qian, ZY., A biodegradable hydrogel system containing curcumin encapsulated in micelles for cutaneous wound healingBiomaterials. 2013 34(27), 6377-6387. doi: 10.1016/j.biomaterials.2013.05.005

[103] Singh, S., From exotic spice to modern drug? *Cell.* 2007;130(5),:765-8. doi: 10.1016/j.cell.2007.08.024.

[104] Madhumathi, K., Kumar, P.S., Abhilash, S., Sreeja, V., Tamura, H., Manzoor, K., Nair, S.V., Jayakumar, R., Development of novel chitin/nanosilver composite scaffolds for wound dressing applications. J. Mater. Sci. Mater. Med. 2010, 21(2), 807-813. doi: 10.1007/s10856-009-3877-z.

[105] Jayakumar, R., Nwe, N., Tokura, S., Tamura, H., Sulfated chitin and chitosan as novel biomaterials, *Int. J. Biol. Macromol. 2007,* 40(3), 175-181. doi: 10.1016/j.ijbiomac.2006.06.021.

[106] Madhumathi, K., Shalumon, K.T., Rani, V.D., Tamura, H., Furuike, T., Selvamurugan, N., Nair, S.V., Jayakumar, R., Wet chemical synthesis of chitosan hydrogel–hydroxyapatite composite membranes for tissue engineering applications. *Int. J. Biol. Macromol.2009,* 45(1) 12-15. doi: 10.1016/j.ijbiomac.2009.03.011.

[107] Niranjana, R., Kaushika, M., Prakasha, J., Venkataprasannaa, K.S., Christy, A., Pannerselvam, B., Venkatasubbu, G.D. Enhanced wound healing by PVA/Chitosan/Curcumin patches: In vitro and in vivo study Colloids and Surfaces B: *Biointerfaces*. 2019, 182, 110339. doi:10.1016/j.colsurfb.2019.06.068

[108] Rachmawati, H., Edityaningrum, CA., Mauludin, R., Molecular inclusion complex of curcumin-beta-cyclodextrin nanoparticle to enhance curcumin skin permeability from hydrophilic matrix gel. *AAPS PharmSciTech*. 2013,14(4), 1303-1312. doi: 10.1208/s12249-013-0023-5.

[109] Rai, M., Pandit, R., Paralikar, P., Nagaonkar, D., Rehman, F. Pharmaceutical Applications of Curcumin-Loaded Nanoparticles. Nanotechnology Applied To Pharmaceu ticalTechnology.*Springer International Publishing AG* 2017, Chapter 6, 139-154.

[110] Strasser, E-M., Wessner, B., Manhart, N., and Roth, E., The relationship between the anti-inflammatory effects of curcumin and cellular glutathione content in myelomonocytic cells. *Biochemical Pharmacology*. 2005 70(4) 552-559. doi: 10.1016/j.bcp.2005.05.030.

[111] Han, S., and Yang, Y., Antimicrobial activity of wool fabric treated with curcumin. *Dyes and Pigments*. 2005,64(2), 157-161. doi:10.1016/j.dyepig.2004.05.008.

[112] Vajragupta, O., Boonchoong, P., Morris, G.M., and Olson, A.J., Active site binding modes of curcumin in HIV-1 protease and integrase. *Bioorganic and Medicinal Chemistry Letters. 2005,*15(14), 3364-3368. doi: 10.1016/j.bmcl.2005.05.032.

[113] Aggarwal, B.B., Kumar, A., and Bharti, A.C., Anticancer potential of curcumin: preclinical and clinical studies. *Anticancer Research*. 2003, 23(1A), 363-398.

[114] Farombi, E.O., and Ekor, M., Curcumin attenuates gentamicin-induced renal oxidative damage in rats, *Food and Chemical Toxicology. 2006*, 44(9), 1443-1448. doi: 10.1016/j.fct.2006.05.005.

[115] Sreepriya, M. and Bali, G., Chemopreventive effects of embelin and curcumin against N-nitrosodiethylamine/phenobarbitalinduced hepato carcinogenesis in Wistar rats *Fitoterapia. 2005, 76(6),* 549-555. doi: 10.1016/j.fitote.2005.04.014.

[116] Panchatcharam, M., Miriyala, S., Gayathri, V.S. and Suguna, L., Curcumin improves wound healing by modulating collagen and decreasing reactive oxygen species. *Molecular and Cellular Biochemistry. 2006*, 290(1-2), 87-96. doi: 10.1007/s11010-006-9170-2.

[117] Phan, T.T., See, P., Lee, S.T. and Chan, S.Y., Protective effects of curcumin against oxidative damage on skin cells in vitro: its implication for wound healing. *Journal of Trauma.* 2001, 51(5), 927-931. doi: 10.1097/00005373-200111000-00017.

[118] Gopinath, D., Ahmed, M.R., Gomathi, K., Chitra, K., Sehgal, P.K. and Jayakumar, R., Dermal wound healing process with curcumin incorporated collagen films. *Biomaterials*. 2004, 25(10), 1911-1917. doi: 10.1016/s0142-9612(03)00625-2.

[119] Dai, M., Zheng, XL., Xu, X., Kong, XY., Li, XY., Guo, G., Luo, F., Zhao, X., QuanWei, Y., and Qian, Z., Chitosan-Alginate Sponge: Preparation and Application in Curcumin Delivery for Dermal Wound Healing in Rat. *Journal of Biomedicine and Biotechnology*. 2009 doi:10.1155/2009/595126.

[120] Epstein, J., Sanderson, I. R. and MacDonald, T. T., Curcumin as a therapeutic agent: The evidence from *in vitro*, animal and human studies. *British journal of nutrition*. 2010,103 (11), 1545-1557. doi: 10.1017/S0007114509993667.

[121] Carroll, C. E., Ellersieck, M. R. and Hyder, S. M., Curcumin inhibits mpa-induced secretion of vegf from t47-d human breast cancer cells. *Menopause*. 2008 15 (3), 570-574. doi: 10.1097/gme.0b013e31814fae5d.

[122] Shpitz, B., Giladi, N., Sagiv, E., Lev-Ari, S., Liberman, E., Kazanov, D. and Arber, N., Celecoxib and curcumin additively inhibit the growth of colorectal cancer in a rat model, *Digestion*,2007, 74 (3-4), 140-144. doi: 10.1159/000098655.

[123] Ikezaki, S., Nishikawa, A., Furukawa, F., Kudo, K., Nakamura, H., Tamura, K. and Mori, H., Chemopreventive effects of curcumin on glandular stomach carcinogenesis induced by n-methyl-n'-nitro-n-nitrosoguanidine and sodium chloride in rats. *Anticancer research*, 2000,21 (5), 3407-3411.

[124] Chuang, S.-E., Cheng, A.-L., Lin, J.-K. and Kuo, M.-L., Inhibition by curcumin of diethylnitrosamine-induced hepatic hyperplasia, inflammation, cellular gene products and cell-cycle-related proteins in rats. *Food and Chemical Toxicology*. 2000, 38(11), 991-995. doi: 10.1016/s0278-6915(00)00101-0.

[125] Azuine, M. A., and Bhide, S. V., Protective single/combined treatment with betel leaf and turmeric against methyl (acetoxymethyl) nitrosamine-induced hamster oral carcinogenesis. *International journal of cancer*. 1992, 51 (3), 412-415. doi: 10.1002/ijc.2910510313.

[126] Usharani, P., Mateen, A., Naidu, M., Raju, Y. and Chandra, N., Effect of ncb-02, atorvastatin and placebo on endothelial function, oxidative stress and inflammatory markers in patients with type 2 diabetes mellitus. *Drugs in R & D*. 2008, 9 (4), 243-250. doi: 10.2165/00126839-200809040-00004.

[127] Pan, R., Qiu, S., Lu, D.-x. and Dong, J., Curcumin improves learning and memory ability and its neuroprotective mechanism in mice. *Chinese Medical Journal (English Edition)*. 2008, 121 (9), 832-839.

[128] Prucksunand, C., Indrasukhsri, B., Leethochawalit, M. and Hungspreugs, K. Phase ii clinical trial on effect of the long turmeric (*Curcuma longa* linn) on healing of peptic ulcer. *Southeast Asian J Trop Med Public Health*, 2001,32(1), 208-215.

[129] Deodhar, S., Sethi, R. and Srimal, R., Preliminary study on antirheumatic activity of curcumin (diferuloyl methane. *Indian journal of medical research*. 2013 138 (1), 170-172.

[130] Heng, M., Song, M., Harker, J. and Heng, M., Drug-induced suppression of phosphorylase kinase activity correlates with resolution of psoriasis as assessed by clinical, histological and immunohistochemical parameters. *British Journal of Dermatology*. 2000, 143

(5), 937-949. doi: 10.1046/j.1365-2133.2000.03767.x.

[131] Balasubramanyam, K., Varier, R. A., Altaf, M., Swaminathan, V., Siddappa, N. B., Ranga, U. and Kundu, T. K., Curcumin, a novel p300/creb-binding protein-specific inhibitor of acetyltransferase, represses the acetylation of histone/nonhistone proteins and histone acetyltransferase-dependent chromatin transcription. *Journal of Biological Chemistry.* 2004 279 (49), 51163-51171. doi: 10.1074/jbc.M409024200.

[132] Aggarwal, B. B. and Harikumar, K. B., Potential therapeutic effects of curcumin, the anti-inflammatory agent, against neurodegenerative, cardiovascular, pulmonary, metabolic, autoimmune and neoplastic diseases, *The international journal of biochemistry & cell biology.* 2009, 41 (1), 40-59. doi: 10.1016/j.biocel.2008.06.010

[133] Basniwal, R. K., Buttar, H. S., Jain, V. and Jain, N., Curcumin nanoparticles: Preparation, characterization, and antimicrobial study. *Journal of agricultural and food chemistry.* 2011, 59 (5), 2056-2061. doi: 10.1021/jf104402t.

[134] Zandi, K., Ramedani, E., Mohammadi, K., Tajbakhsh, S., Deilami, I., Rastian, Z., Fouladvand, M., Yousefi, F. and Farshadpour, F., Evaluation of antiviral activities of curcumin derivatives against hsv-1 in vero cell line. *Natural product communications.* 2010, 5 (12), 1935-1938.

[135] Binion, D. G., Otterson, M. F. and Rafiee, P., Curcumin inhibits vegf-mediated angiogenesis in human intestinal microvascular endothelial cells through cox-2 and mapk inhibition, *Gut.* 2008,57 (11), 1509-1517. doi: 10.1136/gut.2008.152496.

[136] Liang, G., Zhou, H., Wang, Y., Gurley, E. C., Feng, B., Chen, L., Xiao, J., Yang, S. And Li, X., Inhibition of lps-induced production of inflammatory factors in the macrophages by mono-carbonyl analogues of curcumin. *Journal of cellular and molecular medicine.*2009, 13 (9b), 3370-3379. doi: 10.1111/j.1582-4934.2009.00711.x.

[137] Sidhu, G. S., Singh, A. K., Thaloor, D., Banaudha, K. K., Patnaik, G. K., Srimal, R. C. and Maheshwari, R. K., Enhancement of wound healing by curcumin in animals. *Wound Repair and Regeneration,* 1998,6 (2), 167-177. doi: 10.1046/j.1524-475x.1998.60211.x.

[138] Rai, M., Pandit, R., Gaikwad, S., Yadav, A., Gade,A., Potential applications of curcumin and curcumin nanoparticles: from traditional therapeutics to modern nanomedicine. Nanotechnol Rev. 2015, 4(2):161-172. doi:10.1515/ntrev-2015-0001

[139] Saikia, C., Sarmah, M., Das, MK., Ramteke, A., Maji, TK., Biodegradable polymeric nanocomposites: advances in biomedical applications. 2015 Chap. 8, https://doi.org/10.1201/b19314

[140] Tomren, MA., Masson, LM., Hjorth, TH., Studies on curcumin and curcuminoids XXXI. Symmetric and asymmetric curcuminoids: stability, activity and complexation with cyclodextrin. *Int J Pharm*. 2007 338 (1-2):27-34 doi: 10.1016/j.ijpharm.2007.01.013.

[141] Yang, KY., Lin, LC., Tseng, TY., Wang, SC., Tsai, TH., Oral bioavailability of curcumin in rat and the herbal analysis from *Curcuma longa* by LC-MS/MS. *J Chromatogr B Anal Technol Biomed Life Sci*. 2007 853 (1-2):183-189doi: 10.1016/j.jchromb.2007.03.010.

[142] Shutava, TG., Balkundi, SS., Vangala, P., Steffan, JJ., Biqelow, LR., Cardelli, JA., Patrick, O'Neal D., Lvov, YM., Layer-by-layercoatedgelatin nanoparticles as a vehicle for delivery of

natural polyphenols. *ACS Nano*. 2009 3(7):1877-1885. https://doi.org/10.1021/nn900451a

[143] Blecher, K., Nasir, A.,, Friedman A., The growing role of nanotechnology in combating infectious disease. *Virulence*.2011, 2(5),395-401. doi: 10.4161/viru.2.5.17035

[144] Kim, BYS., Rutka, JT., Wchan, WCW., Nanomedicine. *N Engl J Med*. 2010 363(25),2434-2443. doi: 10.1056/NEJMra0912273

[145] Flora, G., Gupta, D., Tiwari, A., Nanocurcumin: a promising therapeutic 610 advancement over native curcumin. *Crit Rev Ther Drug Carrier Syst*. 2013;30(4):331-68. doi: 10.1615/critrevtherdrugcarriersyst.2013007236

[146] Chereddy KK., Coco, R., Memvanga, PB., Ucakar, B., des Rieux, A., Vandermeulen, G., Combined effect of PLGA and curcumin on wound healing activity. *J Control Release*. 2013,171(2),208-15. doi: 10.1016/j.jconrel.2013.07.015.

[147] Bhawana, Basniwal, RK., Buttar, HS., Jain, VK., Jain, N., Curcumin nanoparticles: preparation, characterization, and antimicrobial study. *J Agric Food Chem*. 2011;59(5):2056-61. doi: 10.1021/jf104402t

[148] Krausz, AE., Adler, BL., Cabral, V., Navati, M., Doerner, J., Charafeddine, RA., Curcumin-encapsulated nanoparticles as innovative antimicrobial and wound healing agent. *Nanomedicine*. 2015, 11(1), 195-206. doi: 10.1016/j.nano.2014.09.004.

[149] Jiang, W., Kim, BYS., Rutka, JT., Chan, WCW., Nanoparticle-mediated cellular response is size-dependent. *Nat Nanotechnol*. 2008, 3(3),145-150. doi: 10.1038/nnano.2008.30

[150] Basnet, P., Hussain, H., Tho, I., Skalko-Basnet, N., Liposomal delivery system enhances anti-inflammatory properties of curcumin. *J Pharm Sci*. 2012, 101(2),598-609. doi: 10.1002/jps.22785.

[151] Chirio, D., Gallarate, M., Peira, E., Battaglia, L., Serpe, L., Trotta, M. Formulation of curcumin-loaded solid lipid nanoparticles produced by fatty acids coacervation technique. *J. Microencapsul*. 2011 28(6):537-548. doi: 10.3109/02652048.2011.590615

[152] Dhule, SS., Penfornis, P., Frazier, T., Walker, R., Feldman, J., Tan, G., He, J., Alb, A., John, V., Pochampally, R., Curcumin-loaded -cyclodextrin liposomal nanoparticles as delivery vehicles for osteosarcoma. *NanomedNanotechnolBiol Med*. 2012, 8(4), 440-451. doi: 10.1016/j.nano.2011.07.011

[153] Mohanty, C. and Sahoo, S.K., Curcumin and its topical formulations for wound healing applications. *DRUDIS*. 2017, 22(10), 1582-1592. doi: 10.1016/j.drudis.2017.07.001.

[154] Gopinath, D., Ahmed, M. R., Gomathi, K., Chitra, K., Sehgal, P. K., Jayakumar, R., Dermal wound healing processes with curcumin incorporated collagen films. *Biomaterials*. 2004, 25(10), 1911–1917. doi: 10.1016/s0142-9612(03)00625-2

[155] Gaserod, O., Andersen, PO., Myrvold, R., Gelled biopolymer based foam. FMC Corporation, United States patent application. 2005, publication no. US 2005/0137272 A1.

[156] Mohanty, C., Das, M., and. Sahoo, S.K., Sustained Wound Healing Activity of Curcumin Loaded Oleic Acid Based Polymeric Bandage in a Rat Model. *Mol. Pharmaceutics*. 2012, 9(10), 2801–2811. doi: 10.1021/mp300075u.

[157] Shahnia, M. Formulating curcumin in a biodegradable polymeric composite material: A step towards wound healing applications. 2017, Thesis.

[158] Abdul Khalil, H. P. S., Lai, T. K., Tye, Y. Y., Rizal, S., Chong, E. W. N., Yap, S. W., Hamzah, A. A. ,Nurul Fazita, M. R., Paridah, M.T., A review of extractions of seaweed hydrocolloids: Properties and applications. *eXPRESS Polymer Letters*. 2018, 12(4), 296-317 doi:10.3144/expresspolymlett.2018.27

[159] Sharma, A., Mittal, A., Puri, V. Kumar, P., Singh, I., Curcumin-loaded, alginate–gelatin composite fibers for wound healing applications. *3 Biotech*. 2020,10(11),464. doi: 10.1007/s13205-020-02453-5.

[160] Chiaoprakobkij, N., Suwanmajo, T., Sanchavanakit, N., and Phisalaphong, M. Curcumin-Loaded Bacterial Cellulose/Alginate/Gelatin as AMultifunctional Biopolymer Composite Film. *Molecules*. 2020, 25(17), 3800. doi:10.3390/molecules25173800

[161] Daisy, E.R.A.C., Rajendran, N.K., Houreld, N.N., Marraiki, N., Elgorban, A.M., Rajan, M., Curcumin and Gymnemasylvestre extract loaded grapheme oxidepolyhydroxybutyrate-sodium alginate composite for diabetic wound regeneration. *Reactive and Functional Polymers*. 2020, 154, 104671. doi:10.1016/j.reactfunctpolym.2020.104671

[162] Bergonzi, C., d'Ayala, G.G., Elviri, L., Laurienzo, P, Bandiera, A, Catanzano, O., Alginate/human elastin-like polypeptide composite films with antioxidant properties for potential wound healing application. *International Journal of Biological Macromolecules*. 2020, 164, 586-596. doi:10.1016/j.ijbiomac.2020.07.084

[163] Boztas, AO., Karakuzu, O., Galante, G., Ugar Z., Kocabas F., Synergistic interaction of paclitaxel and curcumin with cyclodextrin polymer complexation in human cancer cells. *Mol Pharm*. 2013,10(7),2676-2683. doi: 10.1021/mp400101k.

[164] Zhao, Y., Dai, C., Wang, Z., Chen, W., Liu, J., Zhuo, R., Yu, A. Huang, S., A novel curcumin-loaded composite dressing facilitates wound healing due to its natural antioxidant effect. *Drug Design, Development and Therapy*. 2019,13, 3269-3280. doi:10.2147/DDDT.S219224

[165] Monica, CGA., Raisa, AMF., Maria DPA., Development and Evaluation of Alginate Membranes with Curcumin-Loaded Nanoparticles for Potential Wound-Healing Applications. *Pharmaceutics*. 2019, 11, 389. doi: 10.3390/pharmaceutics11080389.

[166] Selvia, R.T.1, Prasannaa, A.P.S., Niranjan, R., Kaushik, M., Devasena, T., Kumar, J., Chelliahe, R., Ohe, D.H., Swaminathan, S., Venkatasubbub, G.D., Metal oxide curcumin incorporated polymer patches for wound healing. *Applied Surface Science*. 2018, 449,603-609. doi:10.1016/j.apsusc.2018.01.143

[167] Nostro, PL., Capuzzi, G., Romani, A., Mulinacci, N., Self-assembly and antioxidant properties of octanoyl-6-o-ascorbic acid. *Langmuir*. 2000,16(4),1744-1750. doi:10.1021/la990796j

[168] Nguyen, V.C., Nguyen, V.B., & Hsieh, M.F., Curcumin-Loaded Chitosan/Gelatin Composite Sponge for Wound Healing Application. *International Journal of Polymer Science*. 2013. doi:10.1155/2013/106570

[169] Madhumathi, K., Kumar, P. T. S., Abhilash, S. Development of novel chitin/nanosilver composite scaffolds for wound dressing applications. *Journal of Materials Science*. 2010, 21(2),807-813. doi: 10.1007/s10856-009-3877-z

[170] Friedman, AJ., Han, G., Navati, MS., Chacko, M., Gunther, L., Alfieri, A., Sustained release nitric oxide releasing nanoparticles: characterization of a novel delivery platform based on nitrite containing hydrogel/glass composites. *Nitric Oxide*. 2008,19(1),12-20. doi: 10.1016/j.niox.2008.04.003.

[171] Karri, VV., Kuppusamy, G., Talluri, SV., Mannemala, SS., Kollipara, R., Wadhwani, AD., Mulukutla, S., Raju, KR., Malayandi, R., Curcumin loaded chitosan nanoparticles impregnated into collagen-alginate scaffolds for diabetic wound healing. *Int J BiolMacromol*. 2016 93(Ptb):1519-1529. doi: 10.1016/j.ijbiomac.2016.05.038.

[172] Hegge, A.B., Andersen, T., Melvik, J.E., Bruzell, E., Kristensen, S., Tonnesen, H.H. Formulation and Bacterial Phototoxicity of Curcumin Loaded Alginate Foams for Wound Treatment Applications: Studies on Curcumin and Curcuminoides XLII. *Journal of Pharmaceutical Sciences*. 2011, 100(1), 174-185 doi: 10.1002/jps.22263.

[173] Li, X., Chen S., Zhang, B., Li, M., In situ injectable nano-composite hydrogel composed of curcumin, N,O-carboxymethyl chitosan and oxidized alginate for wound healing application. *Int. J. Pharm*. 2012,437(1-2), 110-119. doi: 10.1016/j.ijpharm.2012.08.001.

[174] Kaur, N., Garg, T., Goyal, AK., Rath, G., Formulation, optimization and evaluation of curcumin-beta-cyclodextrin-loaded sponge for effective drug delivery in thermal burns chemotherapy. *Drug Deliv.* 2016, 23 (7),2245-2254. doi: 10.3109/10717544.2014.963900.

[175] Rachmawati, H., Edityaningrum, CA., Mauludin, R., Molecular inclusion complex of curcumin-beta-cyclodextrin nanoparticle to enhance curcumin skin permeability from hydrophilic matrix gel. *AAPS PharmSciTech*. 2013,14(4),1303-1312. doi: 10.1208/s12249-013-0023-5.

[176] Kiti, K., Suwantong, O., Bilayer wound dressing based on sodium alginate incorporated with curcumin-β-cyclodextrin inclusion complex/chitosan hydrogel. *International Journal of Biological Macromolecules*. 2020, 164, 4113-4124. doi:10.1016/j.ijbiomac.2020.09.013